Smart Guide™
to
Vitamins and Healing Supplements

D1398201

About Smart Guides™

Welcome to Smart Guides. Each Smart Guide is created as a written conversation with a learned friend; a skilled and knowledgeable author guides you through the basics of the subject, selecting out the most important points and skipping over anything that's not essential. Along the way, you'll also find smart inside tips and strategies that distinguish this from other books on the topic.

Within each chapter you'll find a number of recurring features to help you find your way through the information and put it to work for you. Here are the user-friendly elements you'll encounter and what they mean:

The Keys

Each chapter opens by highlighting in overview style the most important concepts in the pages that follow.

Smart Money

Here's where you will learn opinions and recommendations from experts and professionals in the field.

Street Smarts

This feature presents smart ways in which people have dealt with related issues and shares their secrets for success.

Smart Sources

Each of these sidebars points the way to more and authoritative information on the topic, from organizations, corporations, publications, web sites, and more.

Smart Definition

Terminology and key concepts essential to your mastering the subject matter are clearly explained in this feature.

F.Y.I.

Related facts, statistics, and quick points of interest are noted here.

What Matters, What Doesn't

Part of learning something new involves distinguishing the most relevant information from conventional wisdom or myth. This feature helps focus your attention on what really matters.

The Bottom Line

The conclusion to each chapter, here is where the lessons learned in each section are summarized so you can revisit the most essential information of the text.

One of the main objectives of the *Smart Guide to Vitamins and Healing Supplements* is not only to better inform you about vitamins, minerals, and other natural substances, but to show you how to start a supplement program to ensure a lifetime of healthful benefits.

Smart Guide™
to
Vitamins and Healing Supplements

Ruth A. Ricker, Ph.D.

CADER BOOKS

John Wiley & Sons, Inc.

New York • Chichester • Weinheim • Brisbane • Singapore • Toronto

Smart Guide™ is a trademark of John Wiley & Sons, Inc.

The information contained in this book is not intended to serve as a replacement for professional medical advice. Any use of the information in this book is at the reader's discretion. The author and the publisher specifically disclaim any and all liability arising directly or indirectly from the use or application of any information contained in this book. A health-care professional should be consulted regarding your specific situation.

Library of Congress Cataloging-in-Publication Data:
Ricker, Ruth A.
Smart guide to vitamins and healing supplements / Ruth A. Ricker.
p.   cm.   — (The smart guide series)
Includes index.
ISBN 0-471-29633-3
1. Dietary supplements—popular works.   I. Title.   II. Series: Smart guide.
RM217.R53   1998
615'.328—dc21   98-33835

Printed in the United States of America

10 9 8 7 6 5 4 3 2 1

To Alberta Jean Spidell,
without whose encouragement, interest, and assistance
this book would not have gotten written

Acknowledgments

I wish to acknowledge Dr. Eugene Burns, Dr. Alex Cadoux, and all the other medical experts who contributed their knowledge and time to this book. I also want to thank all the people who contributed their anecdotal evidence stories. I especially want to acknowledge Kathleen Meyer for her help in research and chart making. Most of all, thanks to Kate Jaycox who, over the past twenty-five years, patiently answered all my questions about nutrition and showed me how invaluable to one's life a good health food store could be. Finally, thanks to my neighbor John B. Garland for being so helpful with everything.

Contents

Introduction

Healing supplements are in the news: A story on the Fox network news reported that some doctors are treating certain heart and brain disorders with vitamin therapies and seeing positive results. A west coast doctor recently treated an elderly woman in a coma with the antioxidant coenzyme Q10 and revived her. She'd been taken off life support and moved to a hospital where the doctor, who practiced alternative medicine, used this supplement to literally bring her back to life—and she's still alive to talk about it. Air New Zealand distributes herbal blends for passengers to inhale to fight jet lag.

Five years ago, very few doctors practicing traditional medicine would put their careers on the line and openly suggest that *nutritional therapies* have an impact on serious diseases, especially ones affecting the heart and brain. Research into the benefits of antioxidants such as vitamin E was taking off, but coenzyme Q10 and its healing possibilities were not yet available to the consumer market. And commerce, herbs, and aromatherapy rarely met so happily as they do aloft this international airline.

Now more than a trend, the use of healing supplements is part of an important movement focusing on alternate modes of health care. The availability and understanding of healing supplements means that you can make an important difference in the course of your own health care. The *Smart Guide to Vitamins and Healing Supplements* begins to show you how.

If you are just starting out with supplements,

this book offers the most up-to-date research on healing supplements of all kinds—vitamins, antioxidants, herbs, amino acids, and more. Presented in easy-to-understand language is the most essential information that you should know about each category of healing supplements, including:

• What they are;

• Where to find them;

• Which supplements are beneficial for specific conditions—from relieving the symptoms of colds or allergies to alleviating the pains of arthritis to ending premenstrual syndrome to increasing fertility, to mention just a few;

• Safe dosages and how best to take the supplements;

• Cautions to exercise with every supplement, including storage.

If you're seeking answers to certain health issues that can be relieved or alleviated through vitamins, minerals, and healing supplements, then you've turned to the right source. The *Smart Guide to Healing Supplements* will tell you:

• Which form of supplement—whether it's a vitamin, herb, mineral, antioxidant, select homeopathic remedy, or some other—is the most effective when it hits your system, and which forms should be avoided;

• Which healing supplements are best to help specific conditions, from allergies to symptoms of menopause or prostate problems to anti-aging combinations;

• What nutritional supplements *can* and *cannot* do, the best time of day to take supplements for optimum effect, providing you get a go-ahead from a reliable medical practitioner who knows your health history;

• Which supplements are known for their negative effects on health—ones that escalate blood pressure, possibly affect tumor growth, induce extreme mood swings, and other unwanted conditions;

• The most up-to-date research on healing supplements; caveats, or cautions, you must heed on dosages; and how to buy and store supplements.

Here's an overview of just some of what you'll find in every chapter:

Chapter 1 serves as a primer to all the others and offers advice on alternative health care—whom to talk to if your medical doctor will not discuss supplements—how to shop for supplements, and how to understand supplement labels. Chapters 2 and 3 offer detailed looks at the healing power of vitamins and minerals, including the basics, such as the differences between water solubles and fat solubles, how much is too much, and how to get your money's worth buying the right kind of vitamins and minerals.

Antioxidants are explored in chapter 4, and here you'll learn how these amazing nutrients can help decrease physical and mental signs of aging and possibly prevent cancers.

Chapter 5 examines amino acids, the protein components that can provide instant mood eleva-

Be Smart about Healing Supplements

This guide makes no claims that certain substances absolutely *cure* specific diseases or conditions. Neither does this book suggest that any combination of supplements will cause miraculous weight loss or reduce your cholesterol. What this book *does* suggest is that some supplements can promote true healing results—but no supplement takes the shape of a magic bullet.

Healing supplements allow you to be smarter about staying healthy, but be smarter still: *Never* self-treat with any supplement noted in this guide without first consulting your doctor or health-care provider or a licensed nutritional professional who knows your health history and can advise you best on the types of supplements and dosages right for you.

tion, help curb an appetite, and other desirable effects.

The penultimate chapter looks at the healing supplements the plant kingdom has to offer—the most common herbs, roots, algae, and other naturals as well as the best methods of buying, preparing, and taking these types of supplements for maximum effect.

Chapter 7 is a special section on information on supplements and concerns specific to women's health, including herbs that provide the same benefits for women as dangerous hormones, and information on supplements and menopause, osteoporosis, and breast cancer.

In the appendix you'll be able to review the supplements associated with the various physical and mental conditions discussed in each chapter.

Nearly everyone will probably need a healing supplement at some time to make a real difference in getting, and staying, at the peak of health. What's right for you?

The *Smart Guide to Vitamins and Healing Supplements* is the book that can direct you in supplement use—all in terms a beginner can understand and appreciate. This book will be a good, caring, and informed friend that can lead you on your path to an enriched life through better health.

To your very good health!

....

CHAPTER 1

All about Healing Supplements

THE KEYS

• Healing supplements, in all their many forms, are powerful additions to your diet that are used to treat, heal, and prevent diseases and ailments, but they are not "miracle" drugs.

• Advice on using supplements should come from the proper health-care professional, one who knows your concerns and health history.

• You're in control: Make a plan and a schedule for taking supplements to get their maximum benefit.

• Shopping for supplements is like shopping for other necessities: know what to look for, as well as what to *look out* for.

• Deciphering a supplement label will inform you if the product is right for you. The caveats can be explicit or implicit, so read closely and compare.

Is this how you feel now about healing supplements?

You think you could improve your overall health by taking supplements—but don't know which ones and in what amount.

Perhaps your health-conscious friends differ on supplement philosophies: One believes in megadosing on vitamins to heal what ails her and another swears by herbal remedies that sound too exotic. Then, your own medical doctor says it's either pointless or dangerous to take supplements—in fact, he insists, "a good nutritious diet is all anyone needs to remain healthy."

Good health, if you're both smart and lucky, means a longer, better life with strong body systems that can fight for you. Good health shows up in how you look—clear skin, glossy hair, good posture—that healthy radiance.

The truth is: You want to be in control of your health and well-being, so what should you do?

Who should you believe? In fact, what exactly are your friends recommending to you when they recommend "healing supplements," and why does your medical doctor tend to be skeptical of them?

Let's start with a few hints to help you on the healing track.

What *Is* a Healing Supplement?

With the exception of algae, which is said to be a complete food in itself, a healing supplement is *not* a food. It is the nutrient—vitamin, mineral, or amino acid—found in food that is extracted and put into tablet or capsule or liquid form. When taken in certain amounts and combinations, these supplements help heal conditions brought about by dietary deficiencies. Vitamin C, for example, helps heal scurvy, a disease of vitamin-C deficiency. Or the supplement can be the chemical in a plant, such as hypericum in St. John's Wort, that has been found to help heal certain conditions. Thus, the supplements are meant to enrich and enhance the food a person eats, but not replace it.

The Doctor Is In, or, Who's Your Best Adviser?

Most medical doctors will tell you they get an average education of three hours on basic nutrition in medical school—a shocking but true fact. Furthermore, what they learn is more about treating symptoms of malnutrition—for example, poor vision, skin eruptions, bloating—than it is about healing or preventative nutrition.

Although some medical doctors have on their own learned alternate-health practice skills, such as nutritional therapies or acupuncture, most of them simply hold fast to the thing called a balanced diet. "Watch what you eat," they say, "and don't overdo on the fat, sugar, and salt and you'll be healthy." This may be reasonable enough advice, but it assumes everyone eats the same proportion of protein, carbohydrate, and low-fat foods per meal, and has the same biology, energy, and healing needs.

F.Y.I.

Be sure that you never begin a vitamin, nutritional, or healing supplement program without first consulting your doctor or health professional who knows your health history. This ensures the best results for you!

SMART SOURCES

Osteopathic

American Osteopathic
  Association
142 East Ontario St.
Chicago, IL 60611
800-621-1773
www.am-osteo-assn.org

Meanwhile, we know that the University of Arizona Medical College is the first such school in America to include alternative medicine in its curriculum. Hopefully other colleges will follow its lead.

The helping and healing power of supplements can't be ignored, and more mainline *medical* doctors are prescribing supplements to their patients. One New York woman was advised by her internist to take echinacea, a flowering plant commonly found in tablet, drop, or tea form, once a day to strengthen her immune system. A gynecologist concerned about hormone replacement therapy took a step forward with healing supplements. He recommended ginseng to a fifty-five-year-old woman to help stop her hot flashes—an irritating accompaniment for many women during the course of menopause.

These direct recommendations to take such supplements was unheard of in the offices of medical doctors just a few years ago.

If your doctor is attentive to healing supplements and what they can do, you've got a head start—he'll be able to best advise you. If he thinks supplements won't do the job, but you are willing to give them a chance, seek a second opinion.

Other medical practitioners schooled in alternate-health practices and the latest research in vitamins, antioxidants, and herbs are more likely to know what's available out there. And most important, after a consultation, they'll know which supplement is right for the conditions you're seeking to address. Some of these healers get as much training as medical doctors, but with different lines of health-preserving philosophies.

Here are five distinct alternate-health specialists you can talk to. The advantage to consulting with these practitioners is that they tend to spend more time with patients than do medical doctors. The average nontraditional healer spends at least twenty minutes with a patient. Some medical doctors ruled by insurance corporations and managed health-care plans are often forbidden to spend more than ten minutes with each patient.

Osteopaths

Also known as D.O.'s, osteopaths are the closest to medical doctors in that they can legally prescribe medications and do surgery. They also perform spinal manipulations—the "osteo" in their name indicates the underlying philosophy of a strong and healthy skeletal system—and many tend to believe in the value of healing supplements.

To become a licensed D.O.—that is, a doctor of osteopathy—the med student is required to complete four years of basic medical education at an osteopathic medical college. After graduating, D.O.'s can be licensed to open a general practice, or they may go on and choose an area of specialization. This may be, for example, psychiatry, surgery, or pediatrics, for which D.O.'s must enter a two-to-six-year residency program, depending on the specialty.

Chiropractors

These specialists are not legally allowed to prescribe medications, although they are working constantly to change the situation. Chiropractors are the second-largest group of medical care

SMART SOURCES

Chiropractic

American Chiropractic
   Association
1701 Clarendon Blvd.
Arlington, VA 22209
800-986-INFO
703-276-8800
memberinfo@amer-
chiro.org

SMART SOURCES

Naturopathy

American Association
 of Naturopathic
 Physicians
601 Valley St.
Suite 105
Seattle, WA 98109-
 4229
206-298-0125
206-298-0126
www.naturopathic.org
74602.3715@compu-
serve.com

providers, and most people who seek them out first go to them for relief from back pain. A chiropractor's main focus is to keep the body *vital* and free it from any barriers that obstruct its ability to fully experience the life force. Chiropractors use spinal manipulation as the chief source of removing any of the body's so-called obstructions.

Chiropractors graduate from a four-year chiropractic college and, preceding that, complete at least two to four years of undergraduate education—and whether two or four years is sufficient depends on the state's regulations. However, the Council on Chiropractic Education, which provides accreditation to fifteen of the seventeen chiropractic colleges in America, requires that undergraduates seeking to enter their colleges take very specific science courses in, for example, organic and inorganic chemistry and biology. Before opening a practice, chiropractors must pass an exam with the National Board of Chiropractic Examiners.

Naturopaths

These practitioners are also not licensed to prescribe drugs and medications, but their natural, whole-body approach to medical care often allows them to diagnose difficult and chronic conditions that no other doctor's tests can either find or identify. Their emphasis is on disease prevention and the optimization of wellness—and their knowledge of nutrition and healing supplements is vast.

A licensed naturopathic physician, or N.D., attends a four-year graduate-level naturopathic medical school after graduating from a four-year college. Naturopaths are educated in all the basic sciences as is an M.D., *but* they are also schooled in clinical nutrition, acupuncture, homeopathic medicine, botanical medicine, and even psychological counseling to help people change lifestyles. Since all fifty states do not yet certify naturopathic physicians, be sure that any N.D. you consult is licensed and preferably a member of the American Alliance of Naturopathic Physicians.

Homeopaths

Homeopaths cannot prescribe drugs and medications. They base their healing practices on the theory that "like cures like"—the very basis of what makes vaccines work against disease. They give their patients tiny pills that contain minute amounts of a specific substance—related to the symptoms of the patient—which in turn stimulate the person's natural defenses.

Very popular in this country in the early twentieth century, homeopathy was stifled by the growing power of traditional medicine and medical groups. In Europe, one in four French people are ardent believers in homeopathy, and in England, the British royal family follows the practice. Homeopathy continues to flourish and serves millions of believers in India, Mexico, Russia, and other countries.

Interestingly, many practitioners of homeopathy are in fact already M.D.'s, but other health-care providers, such as chiropractors, naturopaths, and even nurses, dentists, and pharmacists, can learn (and have learned) the science in addition to their own specialties. Other than in Connecticut, Nevada, and Arizona, which have state medical boards regulating the practice of

SMART SOURCES

Homeopathy

The National Center for Homeopathy
801 North Fairfax St.
Alexandria, VA 22314
703-548-7790
www.homeopathic.org

International Foundation for Homeopathy
P.O. Box 7
Edmonds, WA 98020-0007
425-776-4147

SMART SOURCES

Acupuncture

National Alliance of
   Acupuncture and
   Oriental Medicine
14637 Starr Road
   Southeast
Olalla, WA 98359
253-851-6895

homeopathic medicine, there's no universally accepted standard determining who may or may not call himself a homeopath.

There are training schools that offer well-respected programs in homeopathy. Be sure to inquire of your homeopath if he's been accredited by the Council on Homeopathic Certification, the American Board of Homeotherapeutics, or the North American Society of Homeopaths.

Homeopathy is making such a strong comeback and uses healing supplements as a matter of course. This Smart Guide takes a closer look at this specialty in chapter 7.

Acupuncturists

Once considered little more than "barefoot medicine," this five-thousand-year-old Chinese healing art has gone mainstream in this country. In fact, a number of doctors with orthodox medical degrees take training in what is considered America's number one alternate-healing art.

The how to and why of it is a bit complicated, but briefly, acupuncturists evaluate patients by testing the main pulses at the wrist and asking patients questions relevant to a complaint. Acupuncture is known to alleviate many disorders with well-placed pricks of special needles, *the* tools of the trade. An able and licensed practitioner in this ancient art is likely to be simpatico to the idea of healing supplements and can likely advise you ably.

Certification in this ancient art requires three years of training at an accredited school for acupuncture and Chinese medicine. These schools are proliferating across the country, and you can locate one by calling the National Alliance of

Acupuncture and Oriental Medicine for recommendations. And be smart! Never seek treatment from someone who has not been certified by a reputable institution.

Supplement Advisers You Should Be Wary Of

Any of these "new age" practitioners should not function as your primary source of advice:

• **Aromatherapy or flower therapists'** very down-to-earth premise is that certain flower, herb, or plant oils can affect a variety of states of mind and physical conditions when they are applied to specific points on the face and the body. Sounds good. "Natural" is not synonymous with "safe," and that which gives off a heavenly fragrance could turn out to wreak havoc on your system.

Most flowers, herbs, fruits, and grasses do have healing, soothing, and nutritional qualities, and such practitioners may be knowledgeable about handling and using these aromatic plants. But unless they have some training in nutrition or the medical arts, they should not be prescribing healing supplements.

• **Massage, Reiki, or acupressure experts** literally employ hands-on techniques, all of which can work like a calmative, muscle relaxant, and in Reiki's case, a supposed restorer of body harmony. There are about eighty or so known touch or massage therapies. Sustained therapeutic touch can feel good, relieve body aches and mental depression, and lower blood pressure, but such practi-

You're in Control: Make a Plan!

When you begin taking supplements, *develop and write down a plan or schedule* and commit to it. A schedule that you stick to prevents you from falling into a cycle where you take supplements in a hit-or-miss pattern, sabotaging your best interests and foiling your intentions. Make changes over time as you need to, increasing or decreasing dosages or adding or substituting others as advised or required. You don't want to stock your refrigerator with supplements on the first day, be enthusiastic and diligent in taking them for the first two weeks, and within a month lose interest and drop them all in the produce bin.

With a written plan, you'll know which supplement to take when, and hopefully work harder to keep the commitment to yourself. Remember, only you are in control of your good health!

tioners, unless they are trained nutritionists, should not advise you on what healing supplements to take or at what dosage.

• **Multilevel marketers in the health supplement business** who are not qualified to advise you on nutritional supplements and doses should also be kept at an arm's distance. Some of these salespeople will be understanding if you want to check with your doctor before ordering their products. Others hope to muscle a sale to get your business and their commission. You can be a pal to a friend in the home-sales vitamin-supplement business, but be prudent first.

• **Infomercials or full-page magazine and newspaper ads** that promise miraculous results if you buy and use their supplements should flash a loud warning bell. No matter how dramatic the claims, never buy what's being advertised without checking with your medical practitioner first.

• **Street-fair vendors, back-of-the-truck salespeople, and swap-meet booths** are great for buying second-hand clothing or household items—not for buying any supplement that could affect your health and well-being. Patronize only reputable businesses for your supplements; get recommendations from your health-care practitioner.

How to Shop for Supplements

If you've decided to augment your diet with supplements, and you've got the time, plan to take one day when you can spend a few hours for comparison shopping. Browse through health food stores, specialty vitamin shops, and pharmacies and supermarkets where supplements are sold. You can determine the best pricing and formulas for the supplements that interest you.

If there's only one health food store, vitamin shop, or drug store in your town, ask to see its suppliers' catalogs and have them order what you need. If you live in a city where you have hundreds of shops to choose from, consider the following criteria before you make your decision:

Know Your Seller

• **The staff should be knowledgeable.** Staffers should know what the store stocks and which brands manufacture the best supplement for the best deal.

F.Y.I.

Nutritional supplements alone do not ensure good health or peak condition. Other than eating a sensible diet, you must *exercise,* which helps heal injuries, build strength and stamina, and make digestion and the synthesis of nutrients more efficient. Exercise as well both stimulates and calms the mind. More important, regular exercise can fight degenerative diseases like heart disease, high blood pressure, and cancer—and it even helps prevent the common cold.

F.Y.I.

Be smart at *home*, too. If you have opened supplements stashed away in an unrefrigerated cupboard or drawer, throw them out. Any oil in capsules can become rancid in a temperature-fluctuating kitchen or bathroom, and vitamins are perishable goods—they can go bad.

• **Shop clerks should not prescribe supplements.** If you ask what's best for, say, easing menstrual cramps, some clerks will tell you what they like and what they "always take, because it works wonders." Such advice is unprofessional, not to mention illegal. And clerks often make recommendations by stating that many customers who come in for relief of menstrual cramps often return for the herb so-and-so in tea or tablet form.

Most health food stores and chain specialty vitamin shops across the country have a shelf of reference books or catalogs to help explain what supplements target what conditions. There you can look up your symptoms (cramps) or health goals (greater energy) and learn what's available. But be sure to check with your health-care professional before buying and taking anything that may not be good for you.

• **The store should be clean.** If the floor is unswept, the shelves unorganized or dusty, and empty cartons or trash line aisles, turn on your heel and shop elsewhere, no matter how charming the owner. Let quality represent quality.

• **Expiration dates should be clearly stamped on every item you buy.** Never buy supplements or food with expired or illegible dates.

• **Be sure your discount vitamin or health food store is reputable.** A good source of money-saving buys, some discount shops lack qualified staff or go out of business so quickly that you can't hold them accountable for defective or inferior products. Some foreign pharmacies have been known to stock knock-off supplements, which may contain a high percentage of fillers, sugar, or oil, and not the nutrients they mention on the label, or their products' ingredients are not properly listed on label.

• **Find a shop that's known to sell or package high-quality supplements under its own label.** Very often these shops develop affiliations with, or are recommended by, a medical doctor, a licensed medical practitioner (such as a chiropractor), or a medical organization. If the shop provides regular guest lecturers or its owner talks smartly about nutrition and healing supplements on local or national radio and TV talk shows, so much the better.

Know What You're Buying

Many consumers tend to pay attention to products that offer great promise—this one will shrink tumors, that one will smooth facial lines, another one lowers blood cholesterol, and yet another kick-starts metabolism and helps you lose weight. Most of us love miracles in a bottle, so buyer beware! Why? The products being touted may be dangerous.

How are you protected? The FDA cannot stop erroneous advertising for health supplements nor can it investigate the companies that produce bad goods until it gets enough complaints about a company's products. (Usually by the time these bogus companies generate enough grievances to warrant an investigation, they are long gone—out of business with no forwarding address.) So the best advice is: *Be smart!* Check the labels for ingredients and dosages, and never be bullied or lulled into buying so-called miracle products from people who have no scruples about taking your money, and maybe making you ill.

How to *Really* Read a Label

All the information you need to know about the product you're buying should be on the label. Below are two examples of typical labels for nutritional supplements—one for niacin, a B vitamin; the other, a B–complex formula—from an imaginary manufacturer.

Each label provides the eleven key points to look for on any label before buying.

SUGGESTED USE: As a dietary supplement, take one (1) tablet daily.
KEEP OUT OF REACH OF CHILDREN.
STORE IN A COOL, DRY PLACE.
Guaranteed:
- *No sugar • No salt • No yeast • Dairy free*
- *No artificial colors, flavors, or preservatives*
- *Labeled for purity and potency • Dated to guarantee freshness.*

EXP. 3/99
MADE IN U.S.A. for Health Central Valley,
Phoenix, AZ
CB3829 JAJ-2

Health Central Valley

YEAST FREE / NATURAL
Niacin

100 mg
100 Capsules

YEAST FREE / VEGETARIAN
Nutritional Analysis:

Each capsule provides:		% RDA*
Niacin	100 mg	500

* Recommended Dietary Allowance for Adults.
In a base of rice bran.

SUGGESTED USE: As a dietary supplement, take one (1) tablet daily.
KEEP OUT OF REACH OF CHILDREN.
STORE IN A COOL, DRY PLACE.
Guaranteed:
- *No sugar • No salt • No yeast • Dairy free*
- *No artificial colors, flavors, or preservatives*
- *Labeled for purity and potency • Dated to guarantee freshness.*

EXP. 3/99
MADE IN U.S.A. for Health Central Valley,
Phoenix, AZ
CB3829 JAJ-3

Health Central Valley

YEAST FREE / NATURAL
ALL B's

Balanced B Complex
100 Tablets

Nutritional Analysis:

Each tablet provides:		% RDA*
Thiamin (Vit. B_1)	50 mg	3.333
Riboflavin (Vit. B_2)	50 mg	29.41
Pyridoxine (Vit. B_6)	50 mg	2500
Cobalamin (Vit. B_{12})	50 mg	833
Biotin (Vit. B_7)	50 mg	17
Choline Bitartrate	50 mg	
Folic Acid (Vit. B_9)	100 mg	25
Inositol	50 mg	
Niacin (Vit. B_3)	50 mg	250
PABA	50 mg.	
(Para Aminobenzoic Acid)		
Vit. B_5 (Pantothenic Acid)	50 mg	500
Lecithin	8 mg	

* Recommended Dietary Allowance for Adults.
In a base of alfalfa, watercress, parsley, and rice bran.

1. If the supplement is *naturally* or synthetically produced. If it is the latter, the manufacturer is not required to say so.

2. If the supplement is in *tablet* or *capsule* form. A tablet is a hard, compressed powder form of a supplement, usually held together with a carrier base of (hopefully) natural and neutral digestible material. The B vitamins sold individually, for example, are likely to come in tablet form.

A capsule is a soft gel pill that contains a more liquid form of the supplement. Vitamin E and lecithin, for example, are usually sold in gel capsules.

3. The *potency* of the supplement per tablet or capsule, measured in milligrams (mg). Some supplements will come in micrograms (mcg).

International Unit (IU) is a standard unit of measurement for the fat-soluble vitamins A, D, E, and K.

4. The *percentage* of the daily RDA, or Recommended Dietary Allowance, of each ingredient in each pill. You won't find this percentage to be uniform for every manufacturer of supplements. In fact, the same manufacturers can change the individual dosages per pill from formula to formula.

For example, looking at the labels note that the 100 milligram niacin tablet provides 500 percent of the RDA, while the 50 milligram niacin dosage in the manufacturer's B complex supplies 250 percent of the RDA.

Always check the label to compare RDA levels of each nutrient, whether from the same or competing manufacturers. If there is no established RDA percentage, there will be a blank space on the label.

F.Y.I.

The Recommended Dietary Allowance (RDA) of vitamins has been frequently updated. This is because research has moved apace over the years, revealing new information about nutrients and supplements. Consumers, too, have come to participate more actively in their own health-care and want to know more about the supplements that they can take.

The RDA has recently been replaced by a daily value measurement, the DRV, or Daily Reference Value. This DRV refers to protein, fat, saturated fat, cholesterol, carbohydrate, fiber, sodium, and potassium. The RDI, or Reference Daily Intake, is the term that will replace RDA and is based on the RDAs of vitamins and minerals. Meanwhile, you will probably see RDA on products for a while.

5. The safe *recommended or suggested dosage* by number of tablets or capsules—or soft gels or teaspoon, whatever the case—per day. Sometimes manufacturers call this serving size.

6. Other ingredients in the pill, or *fillers*—which can give you an idea of the supplement's generally guaranteed nutritional purity. Some supplement manufacturers may use a higher proportion of something like wood or other nonnutritional fillers. The niacin tablet in our example is made in a base of rice bran, while the B-complex capsule is in a base of greens and grain.

7. *Ingredients that are not added* to the pills. Here in the two labels the manufacturer guarantees the products contain no sugar, salt, yeast, dairy products, artificial colors, dyes, flavors, or preservatives.

There are supplement brands that *do* contain artificial dyes, as well as starch, sugar, or yeast fillers. These are usually commercially manufactured multivitamin pills or single vitamin pills that come in typical pill shape or are molded into novelty shapes that appeal to kids.

The list of ingredients—outside the nutritional facts entry—goes by decreasing order by weight. That means that if the supplement lists a filler first—like starch or sugar—it is the

Useful Weight Equivalents

1 microgram	=	1/1,000,000 gram
1,000 micrograms	=	1 milligram
1 milligram	=	1/1,000 gram
1 ounce	=	28.35 grams
3.57 ounces	=	100 grams
0.25 pound	=	113 grams
0.50 pound	=	227 grams
1 pound	=	16 ounces
1 pound	=	453 grams

primary ingredient used in the tablet or capsule, and the filler outweighs the amount of nutritional supplement.

8. The *expiration date.* In this case, it's "3/99," but often the manufacturer will stamp the bottle (rather than the label) and note its presence for consumers. Be sure to check for an expiration date on any product before buying to guarantee potency and freshness.

9. The *warning notice* is a serious caveat. Keep all bottles out of the reach of children. Supplements may look or smell harmless, but they can be dangerous to a hungry or curious child who decides to gobble them down.

10. The *manufacturer's or distributor's name and address* is required by law to appear on the label. If a bottle reads something like Joe's Ultimate Maxi-Power B-Blend, and has no other information about "Joe," don't be tempted by the B-blend's hyped promises or low cost.

11. The *storage recommendation* reminds you to keep the bottle in a cool, dry place to ensure maximum benefit from the supplement.

THE BOTTOM LINE

Healing supplements can address some health issues by relieving, alleviating, or preventing certain physical or mental conditions through vitamin, mineral, herbal, or antioxidant effects.

No matter what type of supplement you might want to take, be armed with information about which form is most effective and which forms are to be avoided. Understand where to buy supplements and from whom to get your okay for taking any supplement. Supplement advice should come only from a licensed health practitioner, not a self-styled nutrition guru. Never take any supplements without first consulting a doctor. Healing supplements aren't for everyone—but a balanced diet and smart nutritional plan *is.*

·····················

The Healing Power of Vitamins

• Vitamins are essential to maintain proper body functions and optimal health. But too much, or too little, can do serious harm.

• These micronutrients are classed into two groups, those that are water soluble and those that are fat soluble, each with advantages and caveats.

• The water-soluble vitamins—the eight Bs and C—are absorbed by the body, then their excesses are eliminated with liquids.

• The fat-soluble vitamins—A, D, E, and K—are absorbed, but their excesses are stored by the body. Caution must be exercised when using these supplements.

• Whether you choose synthetic or natural vitamins is a decision to be made with your health-care practitioner based on your individual circumstances.

itamin B Linked to Reduction of Heart Disease." "Vitamin E Found to Help Prostate Cancer." "Vitamin D May Slow Breast Cancer Rate." These are just a few of the newspaper and magazine headlines touting the amazing discoveries being made about vitamins every day.

Vitamins are usually considered just these mysterious components of food we do not see, enough of which are meant to keep us alive, safe from malnutrition, healthy, and able to propagate the human race. But as healing supplements? How is this possible?

A Basic Vitamin Primer

Vitamins are neither fat, protein, nor carbohydrate, nor do they supply calories. They are, however, something more complex: They're organic molecules necessary for both facilitating important chemical reactions and for building molecules in the body that are vital to health. That is, without vitamins, we'd probably perish from either malnutrition or some disorder directly connected to a vitamin deficiency. Vitamins are found in the foods we consume as well as manufactured in various forms—tablets, capsules, gels, powders, and liquids.

With knowledge of the power of vitamins, we can do a lot more than just prevent malnutrition—we can try to heal physical ailments, too.

The Role of Water and Fat

There are two distinct groups of vitamins:

1. Those that are *water soluble,* such as the eight B vitamins and C; and

2. Those that are *fat soluble,* such as A, D, E, and K.

Water-soluble vitamins can be dissolved in water. The main advantage of water-soluble vitamins is that, except for very huge amounts, excesses of them in the body tend to be washed out in the urine without harming the body. The main disadvantage is that amounts of these vitamins that are needed by the body will be washed out if too much liquid is drunk, or diuretics (substances that prevent water retention in the body) are taken. Caffeine is a diuretic, particularly in the form of coffee. So popping a bunch of water-soluble vitamins with coffee in lieu of food is not a good idea.

Oil- or fat-based vitamins are actually oils. The advantage of these vitamins is that they cannot be dissolved in, or excreted from, the body with water or diuretics. The disadvantage of these vitamins is that in excess amounts not needed by the body they will be stored in the liver. Eventually these accumulations of vitamins A, D, E, and K become toxic and cause unwanted symptoms and effects. Thus their recommended dosages must be strictly observed—more is not better in the case of fat-based vitamins.

A Brief History of Vitamins

In 1906 scientists began the search for substances in food—other than fat, protein, and carbohydrate—that sustained life. When these substances were isolated they proved to be organic molecules. Scientists named them for letters of the alphabet (somewhat haphazardly—with not one but eight vitamin Bs) under the umbrella category of "vitamins," this because of their vital role in keeping all living creatures alive. Eventually, scientists learned how to synthesize these substances into safe, inexpensive supplement pills. By 1925 the first vitamins were on the market.

Because they are *oils,* vitamins A, D, E, and K are measured in IUs, which stands for International Units, instead of metric units of measurement such as grams, milligrams, and micrograms. The world scientific community has agreed upon the IU standard of measurement, so the IU will be the same no matter if you buy any of these vitamins in Belgium, Egypt, or the Philippines, or anywhere else in the world.

How Much Is Enough?

Some nutrition and health experts say you need only the amount of the Recommended Dietary Allowance (RDA) and no more to stay healthy. This school believes that the necessary nutritive intake listed is right enough for nearly every individual, every day.

What is the RDA? The RDA for vitamins and minerals was established by the Food and Nutrition Board in 1941 and has been updated many times over the years. However, the RDA has recently been made obsolete by a newer daily value measurement, the RDI—that is, *Reference Daily Intake.* Expect to see RDI (or DV, for *Daily Value*) on all supplement labels rather than RDAs soon.

Synthetic Versus Natural Vitamins

There are two types of supplement manufacturers—those who follow the commercial route and package synthetic vitamins as well using

Recommended Dietary Allowances (RDAs) for Vitamins, 1989

Age (years)	(RE) Vitamin A	(mcg) Vitamin D	(mg) Vitamin E	(mcg) Vitamin K	(mg) Vitamin C	(mg) Thiamin (B_1)	(mg) Riboflavin (B_2)	(mg equiv) Niacin (B_3)	(mg) Pyridoxine (B_6)	Folic acid (B_9)*	(mcg) Cobalamin (B_{12})
Males											
11–14	1,000	10	10	45	50	1.3	1.5	17	1.7	150	2.0
15–18	1,000	10	10	65	60	1.5	1.8	20	2.0	200	2.0
19–24	1,000	10	10	70	60	1.5	1.7	19	2.0	200	2.0
25–50	1,000	5	10	80	60	1.5	1.7	19	2.0	200	2.0
51+	1,000	5	10	80	60	1.2	1.4	15	2.0	200	2.0
Females											
11–14	800	10	8	45	50	1.1	1.3	15	1.4	150	2.0
15–18	800	10	8	55	60	1.1	1.3	15	1.5	180	2.0
19–24	800	10	8	60	60	1.1	1.3	15	1.6	180	2.0
25–50	800	5	8	65	60	1.1	1.3	15	1.6	180	2.0
51+	800	5	8	65	60	1.0	1.2	13	1.6	180	2.0
Pregnant	800	10	10	65	70	1.5	1.6	17	2.2	400	2.2
Lactating											
1st 6 mos.	1,300	10	12	65	95	1.6	1.8	20	2.1	280	2.6
2nd 6 mos.	1,200	10	11	65	90	1.6	1.7	20	2.1	260	2.6

Note: There are no established RDAs for biotin (vitamin B_7) and pantothenic acid (vitamin B_5). Also, the current RDAs for folic acid for women (nonpregnant, pregnant, and lactating) have been substantially reduced from the previous guidelines set in 1980. This reduction has met with considerable controversy. Consult your health-care practitioner for folic acid advice for your specific needs.

Source: Recommended Dietary Allowances, © 1989 by the National Academy of Sciences, National Academy Press, Washington, D.C.

F.Y.I.

Vita means "life," and *amine,* shortened to *amin,* means "chemical structure."

fillers or dyes in their products, and those who follow the natural market, or those whose products are usually made from natural sources and are free of fillers and dyes. The consumer may favor one type of manufacturer over the other, or may even selectively mix and match products.

Medical doctors, pharmacists, registered dietitians, and those putting out government-regulatory material on nutrition tend to make up a school of thought on vitamins that's generally considered more traditional or commercial. These health experts tend to recommend vitamins made from any source—chemical or natural—that are sold in pharmacies, supermarkets, and other standard retail outlets.

Members of the commercial, or traditional, school recommend you buy vitamins in drug stores and supermarkets, bottled with labels telling you how much of the RDA and/or DV the vitamins provide. As can be seen by the labels, many of these supplements are usually made of synthetic substances and contain fillers.

For instance, both a generic store-brand multivitamin and a fairly well-known retail brand both provide between 50 and 100 percent of most of the vitamins; the generic brand lacked two of the B vitamins, biotin and pantothenic acid, while the well-known brand contained extra minerals: molybdenum, for one.

The ingredients on the generic tablets included sucrose (a sugar, which is listed as the first, and therefore the most plentiful, ingredient) and artificial colors—red dye 40, yellow 6, and blue 1. The well-known brand also contained dyes—not a plus for them.

Those who follow the natural formulas, have a different perspective:

Practitioners who tend to follow this road are usually chiropractors, naturopathic physicians, or homeopathic physicians who believe that natural vitamins are better for maintaining good nutrition, meeting the DVs and *preventing* chronic disease. Their arguments with commercial-brand vitamins, as opposed to natural ones, are these:

• Some fillers, such as sucrose, are said to have all the disadvantages of refined sugar. Even though the amount of such sugars in a pill is small, you don't need it.

• Red and yellow dyes, used for coloring the tablets, are suspected of being carcinogenic.

• According to Dr. Eugene Burns, a naturopath and chiropractor who has been affiliated with the prestigious Swan Clinic in Tucson, Arizona, for more than thirty years, the oil-soluble vitamins, like A, D, and E, should not be taken in a compacted dry powder form. Dr. Burns claims that when processed for tablet form, these vitamins lose almost all their nutritive value. To be most potent, they need to be consumed in an oil base. So their inclusion in a dry multivitamin pill means they may be worthless to you nutritionally.

• Vitamins containing synthetic substances, such as acetate (read your ingredients listing on the label), are not as well utilized by the body as are vitamins made of natural substances.

• To work best, the healing supplements known as antioxidants—such as vitamin E, selenium, or co-enzyme Q10 (more on these in later chapters)—

really should be taken *separately* from the B vitamins. (More on these vitamins in chapter 3.)

Whether you choose the commercial or the natural school is a decision you can make with the aid of a doctor or professional counselor. However, practitioners of both schools they tend to agree on this: Taking a good multivitamin in which all the vitamins are balanced is far better than taking nothing at all.

The B Vitamins

This discussion of the B vitamins begins with a story: Baltimore elementary-school teacher Julie is very attuned to her body's changes, and if she feels off balance or overly fatigued, she has her doctor give her (usually by injection) extra vitamin B complex—her choice of healing supplement. Her reason for taking these vitamins are rooted in her past.

Julie claims that as a child she was an overweight "sickly little slug of a thing." She said, "I had no energy at all and was always a kid's version of depressed and sad." Her mother took her to the doctor, who diagnosed her symptoms and, ruling out a serious disorder, gave her a series of shots of B vitamins. "It changed my life," she said.

Within weeks she lost weight. "I suddenly wanted to be out doing other things besides eating and staring into space!" she said. This twelve-year-old suddenly began enjoying school, making greater efforts to befriend other children, gaining confidence, and, in short, beginning the changes to become the person she is now.

Benefits of the B Group

While the B vitamins serve many purposes, including metabolism of carbohydrates, fats, and proteins; red blood cell development; and the growth and repair of tissues; their main and most immediate benefit is a sense of well-being marked by enthusiasm, energy, clear thinking, and relief from stressed nerves.

This benefit stems from the fact that the Bs keep the nervous system in good repair. Thus, if you have insomnia, vitamin-B supplements may help. (Vitamin B_{12} is known for energy creation.) But if you can't stay awake, paradoxically, they may also help. Biotin, one of the Bs, has been found to help drowsiness. The smart move is to take all the B vitamins in one tablet so you get every benefit and keep your system well balanced.

Other benefits:

• Lowers blood cholesterol levels.

• Lessens symptoms of nutritional deficiency that affect the brain, such as anxiety, hyperactivity, insomnia, mood swings, poor memory recall, and even an inability to learn.

• Helps eliminate hangovers and mitigate the effects of heavy alcohol drinking.

• Alleviates symptoms common to more-serious vascular (vein-related) problems, such as atherosclerosis, the number one killer in America, responsible for about 40 percent of deaths.

• For women, it lessens the chances of cervical dysplasia—a precancerous condition of the cervix—as

F.Y.I.

Keep B-complex vitamins stored in a place that doesn't get direct or continuous light: Light destroys the potency of some of the Bs, especially riboflavin. Your supplement manufacturer packages the Bs in a dark glass bottle to help protect them.

well as birth defects, infertility, and, a serious problem of aging, weakening of the bones leading to osteoporosis.

• Bs can help promote healthy colon function by preventing constipation.

• B complex can help ease the pain of rheumatoid arthritis.

Taking B Vitamins

Never take B vitamins separately without a doctor's supervision. Taking the Bs separately can create serious and harmful imbalances—an excess of one can either neutralize or moderate the good effects of the others—and is too risky for the average supplement user.

B-vitamin supplements must be taken in a B-complex form. For this reason, we are not providing individual dosage recommendations for B vitamins. A reputable B-complex-vitamin formula will contain these individual Bs in proper amounts.

Thiamin (Vitamin B₁)

Thiamin, or vitamin B₁, breaks down carbohydrates in the body and converts them to glucose, the sugar that's responsible for fueling the nervous system and brain. A serious deficiency can cause a disease called beri-beri, which disrupts the nervous system.

Stress increases your need for thiamin, and a lack of thiamin can result in mental confusion.

Research shows that thiamin also helps reduce the chance of developing cataracts—a common and debilitating eye problem for older people.

Riboflavin (Vitamin B₂)

Riboflavin, or vitamin B₂, may help eliminate deficiency symptoms related to red blood cells, nervous system disorders, and poor vision. Studies also show that riboflavin can defuse the toxicity of alcohol and tobacco in the body and help to prevent or lower a person's chances of coming down with cancer of the esophagus.

Niacin (Vitamin B₃)

Niacin, or vitamin B₃, relieves feelings of fatigue by releasing energy from foods. Some doctors are looking to extra supplements of niacin taken in addition to the B complex to also help prevent strokes, improve memory, and enhance mental functioning. Dr. Eugene Burns, for one, says that niacin works by opening the blood vessels in a progression from the base of the neck to the brain. The famed "niacin flush," a prickly, burning sensation experienced by new consumers of niacin, is not, Dr. Burns insists, harmful, just uncomfortable.

Symptoms of niacin deficiency include poor memory.

The B Vitamins and Vegetarians

As the best sources for some of the Bs come from animal products, (particularly in organ meats like liver or kidneys, egg yolks, poultry, fish, fermented cheeses, and dry milk), a lack of B vitamins in the diet could be a problem for vegetarians, especially vegans who eat no animal foods at all, including no dairy foods or eggs.

Especially problematic for vegetarians is the very real possibility of developing a B₁₂ deficiency—the only natural source of B₁₂ is meat—which can lead to severe anemia. Vegetarians usually are prime candidates for B₁₂ supplementation.

Pyridoxine (Vitamin B$_6$)

Pyridoxine, or vitamin B$_6$, helps the body use protein to build tissue, boosts the metabolism of fats, and helps produce red blood cells. Vitamin B$_6$ (and folic acid) is known to help lower plaque (hardened cholesterol) levels and is an immune system aid. B$_6$ deficiencies include symptoms such as anemia, or worse, seizures, and because vitamin B$_6$ is involved with tryptophan production in the brain, insomnia and irritability. If you're on a high-protein diet, your body may require more B$_6$.

Warning! Never take a B complex that exceeds 150 milligrams of B$_6$ a day without a doctor supervision—an overdose of this vitamin can cause neurological damage.

Cobalamin (Vitamin B$_{12}$)

Cobalamin, or vitamin B$_{12}$, works as a team with folic acid (see page 35) in helping reduce symptoms of low energy and can clear mental confusion caused by a deficiency in this vitamin. It has been found to help prevent pernicious anemia and cervical dysplasia, a condition that can turn into cervical cancer. B$_{12}$ can be stored in the body for up to three years.

Biotin (Vitamin B$_7$)

Biotin, or vitamin B$_7$, heals one important deficiency symptom, the inability of the body to absorb the other B vitamins. It also helps break down the accumulation of fatty acids in the body by assisting in their metabolism. Biotin is also, along with vitamins A, B₂, and B₆, great for helping to heal eczema and prevent hair loss. Another benefit is that it strengthens fingernails, a plus for women with nails that are soft or that tend to split.

Pantothenic Acid (Vitamin B₅)

Pantothenic acid, or vitamin B₅, helps reverse some of the deficiency symptoms related to poor growth and development in the body. Some researchers think that increasing the dose of this B can stop or limit hair from graying. Its most positive studies show that it can lower cholesterol—an accumulation of which may cause heart attacks.

Folic Acid (Vitamin B₉)

Folic acid, or vitamin B₉, has a direct effect on lessening depression because it is involved in the production of such brain neurotransmitters as serotonin and dopamine, which help regulate mood, sleep, and appetite. Recent research is also showing that it can help prevent certain heart ailments as well.

This B (which is also known as folate) is so important in preventing birth defects that it is now required *by law* to be added to certain grain-based foods. Folic acid is

"B" Is for Brewer's Yeast

Natural B-vitamin pills are usually derived from yeast. Synthetic Bs, made in the lab from chemicals, are still effective and usually a bit cheaper to buy. And, if you can stand the taste, and do not have a yeast infection, a product called brewer's yeast is absolutely the best source of B vitamins, containing them all in a proper natural balance.

one of the nutrients that are always present in large amounts in supplements made especially for pregnant and lactating women. Be sure to talk to your doctor about this supplement is you are pregnant or planning to have a child.

You're not planning to have a baby in the near future? Rest assured: Since this result of folic acid in preventing birth defects has been proven so conclusively, and accepted by the medical establishment, you can bet other healing and beneficial effects of folic acid are sure to protect the body.

Vitamin C

The RDI or DV—the recommended dose—is a simple 60 milligrams a day, but almost all experts of both the commercial and natural school of vitamin therapy agree that larger doses of the versatile vitamin C are not only safe but preferable. Vitamin C is critical for making collagen (essential for bone maintenance) and is one of the great immune system boosters. Dr. Andrew Weil, a medical doctor and popular advocate of alternative therapies and healing modalities, suggests a dose of 1,000 to 2,000 milligrams three times a day.

Weil accounts for the frequency and safety in taking such doses of vitamin C because it is water soluble. One large dose is not as effective as several smaller doses because the body will excrete whatever amount it cannot use at the time.

Why recommend C as a healing supplement? Vitamin C is a potent antioxidant—actually helping to stop viruses from penetrating cell membranes and causing infections. So can it defuse a few known substances that may cause cancer.

The latest findings on vitamin C also provide good news for problems related to aging: Research shows that this vitamin increases cognitive function in the elderly and relieves sudden mood shifts.

Bioflavonoids

These are a group of compounds found in vegetables and fruits that give them their orange, red, or yellow color and that have been found to increase the potency of vitamin C. The bioflavonoid called rutin, in doses of 200 milligrams, taken with C, helps strengthen capillaries, helps clear up varicose veins and bleeding gums in pregnant women, and can also slow the growth of the common condition called skin tags.

Benefits of Vitamin C

• In combination with bioflavonoids, vitamin C was found to help clear up blood clots in leg veins; this in doses of 1,000 milligrams or more a day.

• It helps alleviate some of the symptoms of hypertension—high blood pressure—when taken in doses of l,000 milligrams or more.

• Further related to cardiovascular health, vitamin C in doses of 2,000 milligrams or more may possibly help prevent heart disease by helping prevent plaque in arteries.

• Vitamin C is crucial in helping to heal wounds.

SMART MONEY

Dr. Jennifer Henkind Ferraro, a board-certified pediatrician in private practice in Stamford, Connecticut, has this advice on vitamin C: "It's generally safe and all right to give a child larger doses if he has a cold, only because C is excreted from the body. If a parent wants to give a child 1,000 milligrams a day for colds, do so only if he can tolerate it well. As far as we know, the only side effect is that too much C may give a child a stomachache.

"Also, avoid megavitamin therapy for kids. Multivitamins come in balanced formulas in preparations that are right for children's needs; for example, in liquid form for toddlers and chewables for children.

"Be especially careful with lipid-soluble vitamins like A and E because they're stored in body fat."

• The formation of carcinogenic nitrosamines in the stomach—nitrosamines come from cured and smoked foods, such as bacon, ham, and hot dogs—may be lessened or defused with C, some studies say.

• Vitamin C has been found to prevent some of the more ravaging effects of chemotherapy, cigarette smoking, and alcohol consumption.

• Doses of 700 to 1,000 milligrams of vitamin C a day has been shown to lower the incidence of a precancerous condition in women called cervical dysplasia.

How Much Vitamin C Is Too Much?

Dr. Andrew Weil says that a limit of 6,000 milligrams spread out over the day, including bioflavonoids, is probably safe and can start you on the way to healing many conditions.

Which kind of vitamin C is right? Dr. Linus Pauling, the late Nobel Prize winner who did pioneering research on this vitamin, said the synthetic form of C was just as good as the natural form.

Other practitioners do not agree. They say, for example, that to do any proper healing, vitamin C must be consumed in a complex containing the natural synergistic ingredients of bioflavonoids, rose hips, hesperidin, and rutin because vitamin C *in foods* is "extremely elusive." What this statement suggests is that by the time orange juice—a top natural source of C—gets to the stores, it's usually lost much of the C content found in the fruit.

Dr. Pauling, one of the first mainstream scientists to espouse vitamin C therapy, claimed he took huge doses of about 10 grams a day. Many alternative-health-care professionals agree with this line of thought, while other more moderate types think 500 to 1,000 milligrams above the minimum daily requirement is quite sufficient.

Warning! Check with your own physician before taking large dosages of vitamin C. A recent study suggested that large doses might actually *cause* cancer. This research has been declared flawed and invalid by many experts, but it could possibly some day be replicated and found to be accurate.

Is Vitamin C Always Good for You?

Some say that large doses of vitamin C creates a tolerance for it by the body. This means that if you've been taking C consistently, you should wean yourself off it gradually. If you stop taking it suddenly, some experts say, you could risk getting sick sooner.

Also, the ascorbic acid in vitamin C could adversely affect some people's kidneys—check with your doctor. If you've had any kidney trouble at all, don't touch vitamin C unless your doctor recommends it.

If there's a known downside to C, it's found in its chewable form: Some of the tablet may stick to tooth surfaces, which may be detrimental to tooth enamel. For this reason, try to brush after eating C chewables.

Vitamin A

The precursor of vitamin A is beta-carotene, which converts into A in the body. Beta-carotene comes from plant sources—such as yellow squash and carrots, all high in this important nutrient. Not as much is known about the effect of taking beta-carotene in supplement form as is known about taking vitamin A in supplement form. One Scandinavian study did suggest that supplementation with beta-carotene might have helped cause cancer in the participants.

Studies performed on vitamin A indicate it is a potent weapon in preventing cancer. However, vitamin A is oil-based and taken from animals. Hence, it is not excreted easily and accumulates in the body. For this reason, very high doses of A can be toxic.

To know the source of your vitamin A supplement, read the label on the bottle. The manufacturer will also tell you if the vitamin A supplement includes beta-carotene and in what percentage. It could look something like this:

Each tablet contains: *% DV*
Vitamin A 5,000 IUs   100% DV
50% as Beta-carotene

This is pretty clear: The tablet contains 100 percent of the Daily Value, of which 50 percent is A and 50 percent is beta-carotene.

If we go back to the commercial and natural ideologies in terms of producing and consuming the vitamin—from either synthetic or from natural sources—we find another argument between the two. The natural group believes that to do the most good, vitamin A should come from natural sources such as fish. Commercial sources, such as the multivitamin makers who manufacture many nationally known and chain-store brands, do not use plant or sea sources but synthetic forms, such as vitamin A acetate.

Benefits of Vitamin A

• Shown to reduce cancerous lesions of the mouth. The result of such a study was reported by the University of Arizona Cancer Center. Vitamin A may also help people who develop oral or esophageal cancer traced to an excess use of tobacco or alcohol.

• Is a major factor in good vision. Vitamin A is essential for the formation of what's called visual purple, which is crucial for night vision. The vitamin also helps heal inflamed eye membranes.

• Is necessary for smoothing rough, dry skin and contains anti-aging qualities. Synthetic derivatives of vitamin A, such as the preparation known as Retin-A, are used topically to treat acne and to reduce wrinkles and lines in facial skin and hands.

• Has an effect on strengthening bones and teeth, and has a connection to preventing the cracking of teeth.

• Keeps the lining of the lungs healthy, say some sources.

• Helps heal wounds and aids in reducing outbreaks of acne.

• Has been found to heal disorders of the stomach lining that lead to ulcers.

• May heal the effects of viruses on the body because of its known antiviral properties.

• Some studies suggest that beta-carotene can protect you from coronary artery disease, which can lead to heart attack, an eventual need for bypass surgery or angioplasty to open badly clogged arteries, and stroke. Beta-carotene, an antioxidant, may lessen the formation of plaque in the arteries.

• Vitamin A has been found to be a most potent killer of free radicals, especially one called singlet oxygen that causes the aging of cells and that may cause cancer. Thus, while vitamin A cannot be said to stop the aging process or to cure cancer, it may help effects of smoking and air pollution on the skin and body.

How Much Vitamin A Is Too Much?

The DV of vitamin A is 5,000 IUs a day for men; 4,000 IUs for women; and 1,600 IUs for children. *Caution:* The body stores vitamin A, so never megadose on this vitamin—it could bring on toxicity symptoms such as nausea, cramps, and flaky skin, among others. Be smart with vitamin A!

Dr. Eugene Burns, a chiropractor and naturopath practicing in Tucson, Arizona, and a member of the natural school, believes that there are occasions when you can take up to 30,000 IUs a day and not suffer any problems. He monitors his patients on such doses—this monitoring is a most critical accompaniment for taking A.

While high doses of beta-carotene are less harmful than vitamin A per se, they can cause the body problems in absorbing other important nutrients in foods. Don't take more beta-carotene or vitamin A than the daily value amount recommended in this book unless under doctor's orders.

In terms of dosages: An average carrot contains 5,000 IUs of beta-carotene. In supplement form, there are no established toxic doses of beta-carotene, but it's wise to never go beyond 15,000 IUs a day.

Warning! Pregnant women should not take high doses of vitamin A because it may cause birth defects.

Vitamin D

Vitamin D, a fat-soluble vitamin, helps the body utilize calcium and can help heal bone deformation due to low levels of calcium and bone loss due to aging. You may reverse the symptoms of vitamin D deficiency—such as severe tooth decay or hearing loss due to a softening of the bones in the inner ear—by taking supplements of it.

By enhancing the body's absorption of calcium, vitamin D promotes growth and health of bones and teeth. More recent and exciting discoveries indicate that a daily exposure to just fifteen minutes of sunlight, a natural source of vitamin D, may be linked to the prevention of breast cancer. Because exposure to too much ultraviolet (UV)

SMART SOURCES

Although almost always associated with milk and milk products, vitamin D is also found in:

Fish oil, especially cod liver oil; egg yolks; salmon; mackerel; tuna; vitamin-D-fortified milk and milk products; bone meal sunlight is linked to skin cancer, getting this small amount of sun is preferable in the earlier or later times of day.

Warning! Both the commercial and natural schools of healing modalities agree that D should not be taken in excess. High doses can damage the liver and other organs.

Benefits of Vitamin D

• It may help alleviate certain respiratory disorders, like tuberculosis.

• Inflammatory bowel disease and conditions in the body that can lead to colon cancers may be relieved by taking the right amount of D a day.

• People who have suffered a stroke or have high blood pressure may heal faster with D.

• The bone diseases osteoarthritis and osteoporosis are strongly linked to vitamin D deficiency. This is because vitamin D is known to aid calcium absorption, and calcium is critical for bone health. Some symptoms of rheumatoid arthritis can also be alleviated with D.

Vitamin E

Vitamin E is not one substance, but a combination of what are known as mixed tocopherols. The most important of these is alpha tocopherol, but the others—known as gamma, delta, beta, and others—are also necessary for activating the full power of this vitamin. The actual form of this vitamin is that of a yellow oil.

Some alternative-health practitioners exalt vitamin E as the most important vitamin of all because it oxygenates the blood and protects cells from the kind of damage that often results in cancer and protects the body from the kind of damage that leads to a rise in bad (LDL) cholesterol and heart disease. For instance, studies have shown that vitamin E was able to protect rodents' mouths from tumor formation, and was able to help protect human blood cells from cancer-causing free-radical damage brought about by smoking cigarettes.

But perhaps the most important function of vitamin E is its ability to block oxidation of fats in the body. Excess fats and cholesterol build up in the healthiest of bodies—and turn deadly when they become rancid, or oxidized. The oxygenating ability of vitamin E appears to slow down and often even halt this fat-oxidation process that can cause many diseases, including cancer.

Benefits of Vitamin E

• Effects of unwise habits like smoking, and unavoidable living conditions, such as living in a city blanketed in polluted air, are particularly mitigated by vitamin E. The oxygen-depriving effects of stress, especially shallow breathing and poor-quality sleep, can be reversed in some cases by vitamin E supplements.

• Studies have shown that 1,200 IUs of vitamin E a day provided a definite postponement of memory loss in Alzheimer's disease patients.

SMART SOURCES

Unfortunately, one would have to ingest nearly unhealthy amounts of these foods in order to get enough vitamin E to help maintain health. Even the most anti-supplement physicians are beginning to admit that supplements are necessary for those who want to take the recommended dose of 400 IUs every day. Here are the best food sources of vitamin E:

Fresh, raw nuts; cold-pressed natural vegetable oils (especially safflower oil); eggs from range-fed chickens; organic wheat germ

Hints for Buying and Using Your Oil-Soluble Supplements

• The decision to choose either the commercial or natural-based supplements is up to you and your doctor.

• Buy your oil-soluble vitamins separately in an oil-base form.

• The best way to get A and D is from nonhydrogenated cod liver oil in a bottle. The second-best source is from cod liver oil in capsules.

• We advise against buying any kind of oil that has been hydrogenated. Hydrogenation is a chemical process that preserves the life of the oil (by making it more saturated) but destroys a great deal of its nutritive value. Experts agree that saturated fats are best avoided. Some say hydrogenation has as much of, if not more than, an adverse effect on the arteries and heart as does animal fat.

• Keep vitamins in the refrigerator, but make sure no moisture or liquids can get into the bottle.

• Vitamin E has been shown to prevent blood clotting of the sort that leads to heart attack; it has also be shown to raise low levels of HDL, the "good" cholesterol.

• Vitamin E has also been shown to prevent and perhaps heal thromboembolisms, dangerous clumping of platelets in the blood. When taken with calcium, this vitamin was found to almost eradicate the risk of fatal pulmonary embolism—blood clots in the lungs—in surgery patients.

• A study with ten volunteers showed that a 2,000-milligram dose of vitamin C taken with 400 IUs of vitamin E was found to cut nitrosamine production up to 95 percent.

• Vitamin E was found to help heal hardened breast cysts in premenstrual women.

• One of vitamin E's qualities that attract its most ardent supporters is its purported anti-aging benefits. While conclusive data is not yet in, many studies with both elderly people and with lab animals show that vitamin E helps protect against the cellular damage that comes with aging. Read more about E and its antioxidant, anti-aging benefits in chapter 4, "The Healing Antioxidants."

How Much Vitamin E Is Too Much?

While the RDI is 30 IUs, many experts, including members of both the commercial and natural schools, claim that at least 400 IUs are needed to provide real healing benefits. Some doctors insist that commercial forms of vitamin E are of little worth since they're made from synthetics that the human body has difficulty absorbing.

The best forms of vitamin E supplements, according to the natural school, are those that say "mixed tocopherols" on the label, along with D-alpha tocopherol. While one study claims the synthetic vitamin E is as beneficial as the natural form, you might prefer the natural form. If so, don't be fooled by any other use of the word, such as "L-alpha tocopherol" or wording such as "alpha tocopherol acetate."

Arizona physician Dr. Eugene Burns, for one, believes the D-alpha tocopherol supplied in a bed of mixed tocopherols mixture is the best form of vitamin E. He recommends that healthy, ath-

SMART DEFINITION

Hydrogenation
A chemical process by which additional hydrogen is added to a fatty chain. This addition preserves the substance and allows it to retain freshness, but the process makes the substance more saturated.

SMART SOURCES

Don't fill up on alfalfa! The fact that one form of vitamin K, vitamin K_1, is derived from alfalfa means a process is involved in that derivation; it does not mean that the body derives this vitamin naturally from that source. The best sources: *Parsley, spinach, kale, and other green leaf vegetables; liver; egg yolks; beans*

letic or very active adults take at least 1,000 IUs a day or more.

One deep-sea diver related that almost all divers take at least 1,200 IUs of vitamin E a day because it increases their stamina and lung capacity marginally.

Dr. Burns suggests that for the most beneficial effect, vitamin E capsules should be *chewed* but not swallowed. Chewing releases the oil contained in the capsules, then you discard the emptied capsules. Yes, natural vitamin Es are more expensive than the synthetic ones, but, claim those in the natural camp, well worth it.

Please consult your doctor before taking this powerful vitamin. Never take vitamin E at all if you are using anticoagulants of any kind to thin your blood, or any other kind of medication that might have a blood-thinning effect.

Vitamin K

Vitamin K is the vitamin that helps blood clot. It's not a vitamin you can go out and buy (nor should you buy it and take it on your own!) but is included here for your information. Vitamin K has three forms—K_1, K_2, and K_3. K_1 is derived from alfalfa; K_2 is produced in the small intestine by bacteria whose function is that of producing this uncommon vitamin; and K_3 is synthetically created. All three are fat soluble and aid the liver's production of prothrombin, a substance vital to the coagulation of blood. In conjunction with this process, vitamin K also creates more calcium-binding sites, which could aid in increased calcium absorption. Your doctor may prescribe vitamin K if you are a "bleeder" (that is, a hemophiliac or one who suffers from some other condition that causes free blood flow). Too much radiation from X rays or other exposure can prevent this vitamin from working in the body.

Benefits of Vitamin K

• Helps with blood-clotting factors.

• It may help older women retain calcium, the loss of which may lead to osteoporosis.

THE BOTTOM LINE

We cannot survive without vitamins. Many of us, though, may not eat a balanced diet. Add to this the daily stresses on the body—air pollution and poor lifestyle habits—and we increase our need for vitamin supplementation to maintain health and prevent ailments.

Be smart about taking the right vitamin supplements: The water-soluble vitamins—the B group and C—can be washed out of the body with liquids. Take them *with food* and avoid liquids for at least a half hour.

The oil-soluble group—A, D, E, and K—must be taken prudently; *never* in megadoses. These vitamins will be stored in the body, especially the liver.

Never take extra vitamins without first consulting your health-care practitioner regarding dosages.

CHAPTER 3

The Healing Power of Minerals

THE KEYS

• Minerals are essential to assisting the body's processes and systems and helping to maintain overall health. But our bodies store minerals, so get advice from a professional before taking individual mineral supplements.

• Multimineral supplements will provide all the major and trace minerals that the body needs, and in the right balance.

• Electrolytes keep the body's cells fueled to perform their important functions, but electrolyte balance can be tipped by liquids.

• Mineral recommended dosages vary by your individual circumstances and the source of the information. Amounts, however, should be supervised by a health or nutrition professional.

• Minerals have varied properties and functions, with experts holding differing views as to their importance.

Minerals are as necessary to health and maintenance of proper body functioning as the all-important vitamins. Even more relevant is the fact that minerals and vitamins are synergistic—that is, they work together, and each needs the other to work optimally. Minerals have also been given much attention recently for their purported healing properties: From the healing aspects of calcium to the stress-combating capabilities of potassium, minerals are continually being studied and explored for their value in fighting and preventing disease and illness.

Mineral Basics: What They Are, Why We Need Them

Minerals are inorganic building blocks that make up nearly every substance on earth, including our bodies. Though the human body is composed of many substances—carbohydrates, proteins, fats, and water, among them—it is the mineral group in our makeup that has the most "lasting" effect: After the life leaves our bodies, and all our organic materials decompose and evaporate, it is the group of minerals that remains. "Ashes to ashes, dust to dust" was never truer.

This is because our bodies absorb and retain minerals, storing them for continual use. Think of vitamins as life-sustaining molecules and minerals as sustainers of the infrastructure systems—such as the oxygen-transport system, the hormone-creation system, the enzyme system needed for absorption of vitamins, to name just a few—that support the processes needed to keep life going. Minerals are first absorbed by the gastrointestinal tract and then released into the blood stream to be used for specific purposes, depending on the mineral.

One smart way to take mineral supplements in the right amount is in a multivitamin-mineral complex that lists minerals—preferably chelated. The more minerals contained in the supplement, the more expensive the product tends to be. Or, a multimineral-complex pill can be purchased separately and taken with vitamins.

No matter what the store clerk urges, and no matter how convincing he or she may be, minerals should not be purchased individually—unless your doctor recommends a specific mineral in single form. As with the B vitamins, minerals can throw one another out of balance if taken in excessive doses.

Major Minerals and Trace Minerals

There are the major minerals—those essential minerals that are found in the human body in quantities greater than 5 grams—and trace minerals, those that are found in the body in less than 5 grams.

Of the more than twenty minerals said to have nutritional value, only seventeen are *vital* to human health. Of these seventeen, seven are major minerals and ten are trace minerals.

Chelated and Colloidal

There are three types of minerals—chelated, nonchelated, and colloidal. Chelated minerals are isolated minerals bound to an amino acid carrier. (*Chele* is from the Latin word "to bind.") Nonchelated minerals are isolated minerals per se. Colloidal minerals are nonisolated in their natural form as found in nature.

Many health-care providers recommend chelated minerals because the amino acid carrier is believed to function like a suit of armor, carrying the mineral more safely to the bloodstream and then being utilized itself, as an amino acid. Nonchelated minerals pose some absorption problems not found with the chelated variety. And most experts agree that chelated minerals are better absorbed by the body.

Other experts have doubts about buying minerals in colloidal form. Colloidal minerals are minerals in tiny solid pieces suspended within a solution obtained from mud in lake beds and other sites of previous volcanic activity. Despite the argument that this mud contains all the nearly one hundred minerals found in nature, the argument stands that many of these minerals are not meant for human ingestion and can even be toxic. Colloidal minerals have, however, been used on farm animals for many years with success and are said by some to be equally beneficial for humans.

With these differing opinions and options—chelated for better absorption, colloidal because it's obtained from its natural state—you may be in a quandry as to what's best. Be certain, the best course to take is the one advised by your nutritionist or physician.

They're listed alphabetically along with their known nutritional values and in what foods or from what sources they are predominantly found. You can read more on each mineral in the mineral profiles later in the chapter.

• **Calcium:** To build bones and teeth; it contributes to good muscle reaction and faster nerve response. Calcium is found in dairy products, cooked bones (for example, in soup stocks or canned salmon), broccoli, and soybean products such as tofu.

• **Chloride:** For brain and liver function and proper cholesterol levels. Found in salted foods and soy sauce.

• **Chromium:** Important for blood-sugar balance, integral to diabetics. Also found to affect cholesterol levels. Found in liver, whole grains, and nuts.

• **Copper:** For iron absorption and to assist in several enzyme functions. Found in shellfish, nuts, cocoa, and poultry.

• **Fluoride:** Known for preventing cavities in the teeth by hardening enamel, fluoride is thought to be important though not generally considered vital for health. It's usually most easily taken in the form of fluoridated water.

• **Iodine:** To help thyroid-hormone production and avert thyroid disorders, such as goiter and cretinism in infants. Iodine is obtained in seafood from the ocean and from food grown in soil with a high-iodide content.

• **Iron:** To avoid anemia, transport oxygen in the blood, and aid in digestion. Iron is found in animal meats, prunes, spinach, and from the cooking process in cast-iron cookware. Cast-iron frying pans are said to be a source of iron; ask your nutritionist to be sure.

• **Magnesium:** For bone and muscle growth; it's required to avoid poor calcium and potassium balance; helps lower the risk of getting osteoporosis as well as possible cardiovascular disease. Find magnesium in soybeans, whole grains, molasses, clams, and some shellfish such as clams and oysters.

• **Manganese:** To promote fertility and lessen chances of sterility; aids in healthy bone and cartilage formation. This mineral is found in rice bran, whole grain cereals, nuts, tea, and ginger.

• **Molybdenum:** For iron storage and because it is found in enzymes. This mineral is found in lentils, sunflower seeds, buckwheat, and lima beans.

• **Phosphorus:** To help strengthen bones and stabilize metabolism. This mineral is found in meat, fish, eggs, and processed foods (from the compounds that are added as preservatives).

• **Potassium:** To help maintain fluid balance and normal blood pressure. Potassium is found in bananas, peanut butter, dried apricots, raisins, and milk.

• **Selenium:** For its antioxidant properties. Found in foods grown in selenium-rich soil and Brazil nuts.

• **Silicon:** To strengthen bones, tissue, hair, and skin. Found in whole grains.

• **Sodium:** Crucial for water balance in the body and good muscle function. Sodium is found in salt, soy sauce, most animal foods, and MSG, the flavor enhancer monosodium glutamate, most commonly added to Chinese food.

• **Sulfur:** Needed by the body for making essential amino acids. This mineral is found in egg yolks, garlic, and high-protein foods.

• **Zinc:** To avoid retarded growth, anemia, poor immunity to disease, and inadequate healing of infection. Find zinc in wheat germ, oatmeal, clams, peas, and some vegetables, depending on the fertility of the soil in which they're grown.

Some experts agree that these seventeen minerals are necessary for good health; others believe that fewer than seventeen are required to maintain a healthy state; yet others recommend a different number or different supplements. For example, you can find multivitamin pills that also contain the minerals nickel, tin, vanadium, and boron, though there's debate as to whether these are required for health and healing.

Authoritative sources stress their own list of minerals as important to health, but they all agree that calcium, magnesium, phosphorus, and sodium are important to a smart nutritional program. Experts also agree that you should get your minerals from nutritional sources and not rely on multimineral supplements to offset a poor diet.

A deficiency of any one of these seventeen minerals can cause a weakening of some body system, depending on the mineral. However, more is *not* better, and minerals should be taken with care, their dosages never exceeded unless recommended by a doctor. Optimal doses for all seventeen minerals, if exceeded, can often be harmful.

STREET SMARTS

Karen, a Louisiana reporter, had been rushing to cover one story after another during the summer in her old car without air-conditioning. By mid-August she felt "different" and was experiencing sudden bouts of anxiety. One hot day, when she was feeling especially phobic, Karen confided in a neighbor, who gave her a glass of tomato juice. Within seconds of drinking it she felt better, "somewhat restored to sanity." She didn't know that the high content of natural potassium in tomatoes was doing its work.

Karen reported her symptoms to her doctor and was told she should have reported the condition immediately. Such mineral deficiencies can turn fatal if not corrected right away.

Minerals to Limit or Avoid

Not all minerals are good for you. Be alert to these:

• **Lead,** which can cause high blood pressure and other conditions. Avoid water (for drinking and washing) that contains lead. A good water filter will help you prevent lead intake. Car exhaust and paint containing lead are also to be avoided. Use a breathing mask if necessary.

• **Cadmium,** which is linked to prostate and other cancers. All cigarette and cigar smoke should be avoided because it includes the contaminan cadmium.

• **Aluminum,** which could enhance Alzheimer's disease. Experts warn against aluminum foil, cookware made of aluminum, and aluminum-containing antiperspirants. Some antacid products, which are often recommended as calcium sources, and talcum powder commonly have aluminum as an ingredient.

Minerals and Electrolytes

To understand the function of the minerals known as electrolytes, it's best first to explain that human beings are actually huge electrical units: The cells of the body function on electrical charges. Electrolytes have positive and negative charges that conduct the body's electricity at a consistent enough rate to keep the body's organs and various systems functioning. Electrolytes control the amount of electricity that goes to the nerves, which allows the muscles to contract, as well as regulating the release of hormones and neurotransmitters.

Electrolyte deficiencies can occur from drinking too little or too much water. The loss of electrolytes through perspiration and dehydration can lead to faintness, loss of nerve function, and, if prolonged, heart arrhythmia, among other problems. Likewise, consuming more than three or four liters of water a day, even in hot weather, will wash enough sodium out of the body to play havoc with the electrolyte balance. Hospitals often treat many patients in the summer who are suffering potentially fatal symptoms of electrolyte depletion due to too much water ingestion.

Electrolytes include sodium, potassium, and chloride. Drinks specially formulated to replace electrolytes are an excellent idea during sports activities, hot weather, and times of great pressure and stress. However, electrolytes should be taken in balanced form—too much sodium increases the need for potassium and throws off the balance.

Mineral Daily Doses

What follows is the Daily Value amounts of minerals adapted from *Prevention Magazine's Health Book*. This list uses two standards of measurement—the newer daily value (DV) or the old U.S. Recommended Daily Allowance (U.S. RDA).

Calcium . . . . DV, 1 gram
Chloride . . . . *No established RDA*
Chromium . . . . . RDA, 50–200 micrograms
Copper . . . . DV, 2.0 milligrams
Fluoride . . . . RDA, 1.5–4.0 milligrams
Iodine . . . . DV, 150 micrograms
Iron . . . . DV, 18 milligrams
Magnesium . . . . . DV, 400 milligrams
Manganese . . . . . RDA, 2.0–5.0 milligrams
Molybdenum . . . RDA 75–250 micrograms
Phosphorus . . . . . DV, 1 gram
Potassium . . . . . . DV, 3,500 milligrams
Selenium . . . . RDA, 70 micrograms for men;
          55 micrograms for women
Silicon . . . . *No established RDA*
Sodium . . . . DV, 2,400 milligrams; RDA,
          500 milligrams
Sulfur . . . . *No established RDA*
Zinc . . . . DV, 15 milligrams

F.Y.I.

Warning! Almost all minerals can do more damage than good if they're taken in mega-doses. Many minerals, in fact, are rendered useless unless taken in combination with other minerals. And yet others are rendered useless *by* other minerals.

Be smart! Remember: Never begin a program of healing mineral supplements unless your doctor okays it. Most health professionals recommend a multimineral complex as the best way to take mineral supplements.

The Mineral Profiles

Calcium

People not getting at least 800 milligrams of calcium a day—ideally, 1,500 milligrams for women in a multimineral pill—may want to consider taking extra calcium in the form of calcium lactate. Simply, this form consists of calcium bound to lactic acid, thus creating an organic form of calcium from an inorganic mineral. Calcium from natural sources, such as oyster shell and dolomite, is not likely to be as potent.

Calcium can help heal certain conditions in the intestines and bowel that lead to cancer of the colon, especially in people with a high genetic predisposition to that disease. Calcium has also been found to cure pica, a disorder that is marked by a craving for dirt, sand, paste, and other nonfood substances.

Taken with vitamin C, calcium is said to help cure colds by increasing the absorption of the vitamin C. Researchers have also connected calcium to relieving symptoms of anxiety, arthritis, depression, high blood pressure, high cholesterol, hyperactivity, insomnia, leg cramps, osteoporosis, periodontal disease, and restless leg complaint.

Chloride

An electrolyte that helps heal nutrient deficiencies by forming hydrochloric acid, a substance that enables absorption of vital nutrients and protein, chlo-

The Calcium-Magnesium Balance: Getting Healthy As You Get Older

The Food and Nutrition Board of the National Academy of Sciences has revised its dietary recommendations for calcium and magnesium intake for men and women *over fifty years old*—and even for people over seventy.

In some cases, the revised dosage is as much as a 50 percent increase over the long-held levels.

These newer recommendations show a change in how researchers are looking at nutrition in terms of preventing chronic disease—the "healing" factor in vitamins and minerals—rather than seeing nutrition in terms of preventing nutritional deficiencies.

Some of the problems that are often inevitable with age may in fact be alleviated with the newer suggestions for daily dietary intake:

• Men and women over fifty should consume at least 1,200 milligrams of calcium a day to prevent a common problem with aging, the loss of bone density. This amount of calcium can be had in the equivalent of four 8-ounce glasses of milk or by taking a calcium supplement.

• Too much calcium isn't wise. The Food and Nutrition Board also advises a daily-intake limit of 2,500 milligrams. The problem: Taking too much calcium can actually contribute to your developing other problems, such as kidney stones.

• For magnesium: Men over fifty should take 420 milligrams daily, and women over fifty need 320 milligrams daily. Magnesium is found in green leafy vegetables, or in a magnesium supplement.

ride is seldom found to be lacking in the diet, except when too many liquids, caffeine, or alcohol is consumed. Its main source is table salt. People who routinely take diuretics should beware of chloride depletion. Deficiency symptoms are twitching muscles, inability to breathe, and finally coma.

F.Y.I.

Chromium picolinate is thought to help the body absorb chromium more readily. But its most popular claim is that it helps promote weight loss and build muscle. However, researchers at Dartmouth College discovered that chromium picolinate can accumulate in the body and eventually cause cancer. Be wary of the those miracle weight-loss drinks in a bottle. There's no substantive evidence to support their promise.

Chromium

Chromium is a mineral once thought to be toxic. Research in the 1950s revealed chromium to be the active component of the Glucose Tolerance Factor—the substance in yeast that stabilizes blood sugar. Thus, chromium can help conditions associated with poor blood-sugar balance, such as diabetes. It can also help lower cholesterol levels.

Despite claims by natural-foods people and bodybuilders that this mineral causes weight loss by speeding up the metabolism, some experts, such as Dean Edell, M.D., reported that a high dosage of chromium was found to cause cancer and chromosomal damage in animals. So be advised: Do not take extra doses of chromium other than the amount contained in a multimineral supplement (do not exceed 300 micrograms a day).

Copper

Copper in the body as a healing mineral? Yes, say some experts. When it is combined with zinc, this mineral has been found to heal by fighting free radicals.

Copper is used to help heal anemia because it helps transport iron and form hemoglobin, the red-blood-cell component that takes oxygen from the lungs to all other parts of the body. Copper can also lessen the effects of stress on nerves because it helps maintain the covering of nerves, called the myelin sheath.

Copper is also an antioxidant. Unbound from proteins and enzymes, however, copper is a free radical in the body that may cause cancer. Too much zinc can cause a deficiency of copper. It is especially important that this mineral be taken in proper balance with zinc and other minerals. A multimineral supplement is best.

Fluoride

This mineral prevents cavities in the teeth by hardening enamel so bacteria cannot penetrate the tooth and by attacking the bacteria themselves. Too much fluoride can mottle the teeth, however, and cause gastroenteritis, among other conditions.

Evidence also suggests that fluoride increases cancer risk. In any case, no one should take more than 4 milligrams a day and never more than 1 milligram per liter of water, which is 1 part per million. Four parts per million of fluoride, or 4 milligrams per liter, is said to be toxic.

Iodine

One of the founding mothers of popularizing the importance of diet and nutritional therapies, Adelle Davis devoted her life to nutrition because of her own problems with iodine deficiency. She grew up in the midwest in the 1930s, ate no fish, and lived in an area of iodine-poor soil. She suffered from goiter—a disease of the thyroid gland and a terrible complication of iodine deficiency. Another symptom of iodine-poor diet is sluggish metabolism.

Iodine can prevent cretinism in babies, which,

Too much iron has been found to cause cancer in men. The use of iron supplements is never a good idea— unless they are prescribed and their use is monitored by a physician.

if untreated, leads to retardation of mental and physical development.

Iodine is readily available in seafood from ocean sources—rather than fish harvested from freshwater lakes or rivers. Although iodine is added to salt—you can buy a very inexpensive box of iodized salt in any supermarket—salt may increase your sodium intake and raise your blood pressure. Two servings a week of a low-fat ocean fish like halibut or mackerel can provide your iodine needs and help boost thyroid function naturally, without risk of increased salt intake.

The Daily Value (DV) of iodine for good health is 150 micrograms.

Iron

Many experts recommend taking iron-free mineral supplements for the following reasons:

• Iron in certain forms (especially in a synthetic form such as ferrous sulfonate) can cause constipation because the body has a hard time breaking down these forms of iron.

• Iron interferes with the absorption of vitamin E.

• Iron absorption is increased by vitamin C. Thus, people taking a lot of C plus extra iron could be absorbing too much iron. Iron overload, which could be any amount over the Daily Value (DV) of 18 milligrams, has been found to increase the risk of heart attack and cancer.

However, iron is very important, especially for women—and iron obtained from plant sources is not constipating. Ferrous fulmerate or ferrous peptonate are two examples of such a form. The best rule is: Don't take extra iron of any kind without asking your doctor first.

Iron was the first mineral to be claimed as vital to life. It heals anemia, which is a condition caused usually by a lack of iron. This iron deficiency produces such symptoms as fatigue, slow mental development, dejectedness, and feelings of anxiety and tension.

Premenopausal and pregnant women are likely to be most deficient in iron because pregnancy and blood loss—which occurs during menstruation—deplete stores of iron. In a form of anemia known as iron-deficiency anemia, red blood cells have too little hemoglobin and thus carry too little oxygen for the needs of the body. Some symptoms of this iron deficiency are diminished energy and lethargy.

Iron absorption in the body is inhibited by antibiotics, antacids, and calcium in carbonate form, but chelated minerals and vitamin C help absorption. Once absorbed, iron is not excreted, which is why too much iron in supplement form builds up and cause problems. Too much iron may even cause certain types of cancer, infections, and heart disease.

Adult men and women should get 18 milligrams of iron a day; women who are pregnant need 30 milligrams a day; and those who are lactating should get 15 milligrams a day.

Magnesium

This mineral can help heal cardiovascular illness, chronic fatigue syndrome, and muscle cramps. It can also help prevent kidney stones. If you take too much magnesium, it can also cause calcium deficiency—so the two minerals should always be taken in balance.

Large amounts of magnesium are present in laxatives and antacids, so accidental overdosing is a possibility.

Manganese

This mineral is known to help heal inflammatory diseases, osteoporosis, sprains and strains. Manganese provides a good example of why minerals must be taken in a balanced complex: In too high doses it may interfere with neurotransmitter functioning in the brain. Too much extra iron causes manganese to be excreted; high doses of other minerals can keep manganese from being absorbed. Its best food sources are whole wheat flour, peas, and brown rice.

Molybdenum

This mineral may help heal cancer of the esophagus, a recently discovered symptom of deficiency of this mineral. Molybdenum is now thought to have a detoxifying effect in the body: New research is showing that it can help mitigate the harmful effects of alcohol and environmental pollutants. Molybdenum also shows promise in helping treat asthma and preventing the formation of carcinogenic nitrosamines. Too much molybdenum can decrease copper levels. Its best food sources are lamb, grains, nuts, lentils, and squash.

Phosphorus

Phosphorus is one of those minerals whose balance in the body is easily tipped one way or the other. Phosphorus deficiencies can be caused by taking too much iron, magnesium, and aluminum—the last two of which are abundant in many antacids, used for relieving stomach and intestinal distress.

An excess of phosphorus can occur if you eat too many processed foods and drink too much soda—both contain phosphorus compounds, such as phosphoric acid. In fact, too much phosphorus can create imbalances or deficiencies of magnesium and calcium if not taken in strict 1:1 balance with calcium.

Phosphorus should not be taken as a single supplement but as part of a balanced mineral supplement to heal the symptoms of phosphorus deficiency such as fatigue, general malaise, feelings of body weakness, and a shortened attention span.

Potassium

Potassium is unique in that it is an electrolyte, meaning it's capable of conducting electricity, and is crucial for brain function.

However, serious injuries and prolonged stress can counteract the effectiveness of this mineral.

Potassium's most heroic healing role is that of preventing heart attacks caused by potassium deficiency. Potassium, along with sodium and chloride, helps heal effects of stress and plays a significant role in preventing the body from going into shock when wounded.

WHAT MATTERS, WHAT DOESN'T

What Matters

• Minerals and vitamins work together. Opt for a supplement that includes both.

• Avoid taking minerals individually, unless supervised by your health-care practitioner. Multimineral complexes will give you all the minerals, and in the right portions, that you should need.

• Get some of your minerals naturally, from food sources.

What Doesn't

• "Miracle" minerals advertised in magazines and fliers.

• What other people or store salespeople recommend.

• Concern with all the many different minerals on the store shelf; look at the multi-complexes first.

It can also reverse confusion brought about by excessive perspiration, and relieve cramping, fatigue, nausea, vomiting, and increased urination, also caused by potassium loss. Caffeine, diuretics, and laxatives can increase excretion of potassium and result in deficiency.

When you see athletes slugging down Gatorade or a similar energy drink during timeout, it is usually because the drinks contain a significant source of potassium to make a difference in their energy levels for continuing the game. Athletes sweat out crucial minerals. The same is true of anyone sweating in the heat of day.

Because this mineral regulates fluids in the body, it is an important healing supplement for people suffering the effects of heat exhaustion, such as the effects of too much perspiring.

A doctor should be consulted about taking potassium supplements.

Selenium

Think of heavy metal not as a crash of highly amplified guitars and flashing strobe lights, but the clash of clean air and lead-based toxins. Lead is everywhere—belched into the air by exhaust from leaded gasoline—as well as found in many common household products, such as some paints and can seals.

The more lead and other heavy metals, such as mercury, there are in your environment, the more you need to be detoxified. Which detoxifier? Studies show that selenium seems to lower the risks associated with the accumulation of such heavy-metal toxins.

The healing properties of this mineral seem to

be authenticated. For one, higher amounts of it have been found in the bodies of people who *don't* have cancer than in those who do. In a comparison study of 101 patients with melanoma (a form of skin cancer) and 57 people free of melanoma, it was found that the selenium was much lower in the cancer group than in the healthy group. A similar study in Finland revealed higher amounts of selenium in men who didn't have cancer than in those who did.

Selenium can also help heal the following symptoms of its deficiency: infertility, chronic fatigue, and heart muscle disease.

Amounts beyond 200 micrograms a day can cause loss of hair and deterioration of nails. Selenium is potent as a healer, but it can be toxic in high doses. Don't take more than is contained in your multivitamin-mineral pill. (There's more on selenium in chapter 4, "The Healing Antioxidants.")

Silicon

Silicon is found mainly in sand, quartz, agate, and flint. However, it is also a nutrient to consider in terms of good health. Signs of silicon deficiency include brittle nails, dry hair, and blotchy skin. Silicon may help heal atherosclerosis—the narrowing of veins and capillaries due to plaque—and aid in lessening the damage from osteoporosis.

Too much silicon—above 50 milligrams a day—may lead to Alzheimer's disease. In fact, along with aluminum, traces of it have

Checking Out Vanadium

The little-known mineral vanadium has been found to prevent cancer in mice. More recently, vanadium has been suspected of possibly being a factor in controlling obesity, one symptom of vanadium deficiency.

been found in the brains of those afflicted with this disease.

Sodium

There's much talk about this electrolyte in relation to hypertension—that is, its *possibly* being a factor in high blood pressure—and its connection to what is typically a high-salt diet. Salt, or sodium chloride, is one of America's favorite tastes—and we eat upward of two to five times the recommended limits per day!

And sodium's benefits in the body?

Sodium is an important mineral that can reverse mental confusion brought about by excessive perspiration, as well as muscle weakness, dizziness, decreased blood pressure, increased heart rate, and shock from sudden sodium loss due to diarrhea, vomiting, and burns.

It's important to keep your sodium intake at a moderate level and to keep the electrolyte group in the proper balance.

Sulfur

This mineral can heal the buildup of toxic substances in the body by binding with and eliminating them. Enough sulfur can change a sallow complexion by putting a glow in it, and it may also increase low energy. Sulfur springs—water high in sulfur content—have been said to help heal the pain of arthritis. As part of many amino acids, sulfur is crucial to the structural health of the body. It is usually present in mineral-complex supplements. Sulfur's best food sources are broccoli and other cruciferous vegetables. Lean beef is also a good course, as are beans. Deficiencies of sulfur are uncommon.

Zinc

Considered by some to be one of the latest wonder minerals, zinc can, for some, strengthen the immune system, heal infections, and, most of all, fight colds and sore throats. Zinc can be especially helpful if taken when cold symptoms first appear and taken in concert with vitamin C.

This mineral helps heal wounds—cuts and scrapes—and ulcers much more quickly than they would heal without it. Zinc can prevent, but not heal, the most dramatic symptoms of its deficiency—dwarfism and undeveloped genitals.

The most critical symptoms of zinc deficiencies—frequent illnesses due to a poor immune system—can often be relieved, or with some takers reversed, with zinc supplements or zinc-rich foods, such as wheat germ, oysters, and peas. Zinc has also been found to help heal acne and eczema. You can find topical ointments to apply directly to the skin at most drug stores.

Some studies show that zinc may help pull the starving anorexic out of her or his rejection of food and nutrients so she or he can begin eating again. Zinc has also been shown to help relieve benign prostate problems, herpes infections, and intestinal ulcers and infections.

Results of clinical studies proving the cold-fighting powers of zinc are noted on some packages of zinc-containing cough drops. However, taking too much zinc is truly overkill—overdosing

F.Y.I.

Do not take more than 50 milligrams a day of zinc. An overdose can eventually lower the levels of HDL—the good cholesterol—in the body. Get your zinc in a good multivitamin-mineral pill.

If you are coming down with a cold, lozenges or cough drops containing 11 milligrams of zinc can be taken for up to five times a day until your cold symptoms are relieved. Then follow the 50-milligrams guideline.

on zinc can cause serious problems, even cancer. For this reason, extra zinc is recommended only in the event of a cold.

Zinc taken in amounts of 100 milligrams a day or more can raise the bad cholesterol levels in the body and thus contribute to heart attack.

THE BOTTOM LINE

Minerals are critical to health—they are essential for the growth and mainte-nance of bones and teeth; necessary for heart function and transporting oxygen in the blood; integral to regulating energy output; and all-important to brain functioning, even affecting our mental outlook.

Nature, however, has created a delicate balance between the minerals in the body. So don't overdo, hoping to replace what you think is deficient, or overdose, hoping to cure an ailment. Always consult your medical practitioner about which mineral supplement is right for you before embarking on a supplement program.

The Healing Anti-oxidants

THE KEYS

• Antioxidants are valuable metabolic disease fighters that strengthen the body's cells and prevent illnesses.

• Free radicals, unstable by-products of normal cellular processes, can cause bodily damage, especially when combined with environmental and lifestyle toxins.

• The normal process of aging brings about destructive changes that can be addressed by antioxidants and medical supervision.

• Allergies are a response of the immune system that can be helped with antioxidants.

• The healing properties of antioxidants have been shown to dramatically affect cancer treatment and prevention.

• Adopting an antioxidant regimen, under a health professional's advice, is a wise step in preventive care.

One of the latest discoveries of science, antioxidants are those supplements—vitamins and minerals, plants and herbs, and, in some cases, amino acids—that work together in certain combinations to do an especially efficient job of fighting the dangerous carcinogenic and immune-system-damaging effects of free radicals from normal cell metabolism. This chapter not only advises which supplements are the best antioxidants, but explains why they are so necessary for dealing with the many natural and environmental threats to our bodies.

Antioxidants and Their Importance

Antioxidants knock out viral and bacterial infections, strengthening the cells that activate the immune system to resist invaders. Some antioxidants, such as garlic and vitamin C, contain antiviral and antibacterial agents themselves. Other antioxidants protect against what may be a fatal illness—like cancer—or fight the inevitability of aging and the breakdown of tissues and bodily functions.

Free Radicals

Some of the threats our bodies fight every day are in the form of *free radicals*. What are they?

A free radical is a waste product produced by

normal oxygen metabolism that goes on in the cells. These waste products seek attachment to healthy cells and in making that attachment, the free radicals damage the healthy cells' nuclei, including their DNA patterns. Also, just like vampires who make of their victims *other* vampires just by sinking fangs into their victims' necks, the attacking free radicals convert their victim cells into new free radicals! Eventually, there are too few healthy cells left in the body to ward off cancer and other diseases and maintain bodily health.

Our bodies are under constant threat from agents of minor or major destruction in the body from free radicals. Experts think that free radical damage is the cause of at least sixty illnesses, including cancer and heart disease.

Free radicals are also produced by the body's attempt to metabolize pollutants. Pollutants in the air, X-rays, and cigarette smoke—both inhaled and exhaled—are just three of the stressors on the body's cells that result in the production of free radicals.

How Antioxidants Fight Free Radicals

While free radicals cause certain breakdowns in body tissues, an *antioxidant* can prevent or counteract damage by the unstable oxygen molecules. Vitamins C and E, carotenoids like beta-carotene, and minerals like selenium, zinc, and glutathione, an amino acid, are all powerful antioxidants.

By taking certain vitamins and minerals, you can create an *antioxidant* defense system that works with antioxidant enzymes already in the

What's in a Head of Garlic?

Garlic is being viewed more and more as a miracle plant. Besides being a natural antibiotic, garlic, which contains flavonoids, is an excellent antioxidant. Garlic supplements you find bottled in vitamin shops or health food stores can be expensive, so your best bet is to eat *natural* garlic. Cooked garlic is almost as potent a healer as raw garlic—which you can, for example, chop into salad dressing. Follow up with a sprig of parsley or a chlorophyll tablet to help do away with garlic breath.

Cooking garlic is simple: Lightly oil a head of garlic, place in a garlic roaster—usually a small four-inch-round crock pot with a lid—or wrap it in microwavable plastic wrap and cook it for about two minutes in the microwave. Or oil a cookie sheet, oil the garlic head, wrap it loosely in tin foil and roast for half an hour in the oven. The head is done when the cloves are soft and can be squeezed from the skins.

Garlic is a delicious and very nutritious accompaniment to any meal and it's likely to destroy any germs or fungi picked up by the body during the day.

body to destroy these free radicals.

Before talking about the nutrients needed for this life-enhancing and even life-extending antioxidant effect, let's examine some of the increasing threats to Americans from free radical damage.

Why the Free Radical Problem Is Worse Than Ever Before

Part of the problem of free radical formation is a result of the modern age, most notably, pollution, pesticides, smoking, and more.

Before the Industrial Revolution, food was

grown and harvested and, depending on the quality of the soil, full of vitamins and minerals in balance to keep people healthy.

Then came the rise of factories, and with them, pollution in the air and water, and the growth of densely populated cities. By the mid-twentieth century, cars and their exhaust were a serious factor in pollution problems. Many foods were stripped of most nutrients in the processing for mass consumption; refined sugar, chemical additives and caffeine abounded in most supermarket products. Hundreds of chemicals were used in housing materials, including formaldehyde and asbestos.

Drinking water was chlorinated and infiltrated by pesticides used in agriculture and landscaping. Aerosol sprays and air conditioners created further problems with air quality and damage to the ozone layer. Radiation from appliances, computers, digital clocks, nuclear energy facilities, and cigarette smoke didn't make the atmosphere cleaner or clearer. Plant foods tended to wear a film of pesticides used by growers.

All this exposure to pollution still results in free radical production that most human bodies cannot handle without help. Cancer, heart disease, atherosclerosis, and other potentially fatal diseases as well as learning disabilities and emotional problems have increased dramatically in the past few decades. Statistics aren't quite in on the increase of debilitating autoimmune diseases and allergic reactions—but these complications, too, have proliferated.

Fighting Back

These conditions of widespread allergic reactions to a myriad of substances in the environment are becoming more common. Anyone who doubts the dangers of environmental pollution is advised to rent the independently made film *Safe,* now on video. *Safe* tells the fictional, but very realistic, story of a healthy, beautiful young upper-middle-class wife in Los Angeles who suddenly suffers multiple allergic reactions to environmental chemicals in car exhaust, ordinary food, plant pesticides, and in the very air she breathes. Her only means of survival is going to live in a spartan mountain retreat operated for people with this allergic condition.

No one can say for sure if an antioxidant regime would have kept this woman and others like her healthy. But it is known from many clinical studies that antioxidants help the body fight off incapacitating reactions to our ever more polluted world.

More than ever, we need antioxidants to help us feel good and feel better.

Antioxidants and Aging

One reason why humans begin to see the aging process begin to go full bloom at around the age of fifty is, truth be told, aging is programmed into our genes. Mother Nature, as old as the species itself, phases out members of the human race once their childbearing years are over, and makes room for new generations.

Nature has not caught up with medical, technological, scientific, and attitudinal progress of the modern world, all of which are intent on defying these natural forces. Can man and woman routinely live to one hundred years or more? Antioxidants, some say, could be an important factor in determining longevity.

Until the nonaging or deferred-aging process becomes a reality, all of us will age in a generally similar way—a slow-down of energy, slight or severe compression of the bones in the spine, graying hair or hair loss, thickening of the abdominal area, slower repairing of wounds, lines, wrinkles, and age spots. Others cope with hearing loss, circulatory and gastrointestinal problems while still other manage with high blood pressure or memory loss. And more.

Another consequence of age is that older people can no longer *absorb* vitamins as well as they could when they were younger. This change in how the body processes the food it needs to survive means that the aging body cannot get as many nutrients from food—even from a healthy diet.

As you age, be smart and avoid overcooked or starch-laden comfort foods, such as white rolls, mashed potatoes, and pancakes, fattier cuts of meat and sugary desserts and snacks that have few nutrients at all.

One of the most serious problems in aging, though, is damage to or destruction of the immune system by free-radical damage. For some, this process happens quickly via some fatal disease. With others, the erosion of the immune system is slower and does not actually wear down until ripe old age. But this condition we call age can be curtailed, slowed down, or in

F.Y.I.

Reduced incidences of macular degeneration, one of the prominent causes of age-related blindness, have been reported in studies of people whose diets are high in the antioxidant carotenoids from vegetables and fruits.

some cases, reversed, to a great extent with the use of antioxidants.

How Antioxidants Can Help

A routine regimen of the three key antioxidants—vitamins C, E, and A—plus beta-carotene and selenium augmented with glutathione or glutamine can help heal damaged cells by eliminating free radicals found in fatty foods.

Yet another problem of aging that antioxidants alleviate is the decreased ability of the body to get oxygen to the cells through the lungs. Coenzyme Q10 and vitamin E and selenium are believed to be especially effective in helping oxygenate the body by dilating—that is, opening—the small blood vessels and capillaries, thereby stabilizing circulation and reducing the risk of vascular disease.

Antioxidants and the Mental Problems with Aging

Depression, mood swings, and mental incapacity increase as people get older and are not just "in the mind." In many cases, biology plays a role.

The fats in the fatty tissue of the brain eventually deteriorate as you age—a process called lipid peroxidation, thus giving rise to senility and possibly even Alzheimer's disease. Experiments and studies with both vitamin E and selenium have been found to reduce the amount of deterioration to brain tissue.

There are many ideologies about supplements

used for the aging body. One such program comes from Roy Walford, M.D., an expert on aging. He's found that a reduced calorie diet of 1,500 to 2,000 calories a day along with a daily regimen of antioxidants delay aging symptoms. He suggests taking:

Vitamin E: 300 IUs

Vitamin C: 1,000 milligrams

Selenium: 100 micrograms

Coenzyme Q10: 30 milligrams

These amounts are considerably less than those recommended by other experts—some of whom might increase the dosage by three to five times. The reader is advised to make up his own mind about dosage amounts with the aid of a doctor or nutritional advisor's okay.

Antioxidants and Allergies

Allergies are the result of overreactions to any number of substances—like pollen or perfumed room spray or wool—that, in a way, jolt the immune system.

The immune system is meant to ward off any threats to health, but it can often confuse non-threatening substances that are ingested, inhaled, or touched with those, such as bacteria and viruses, that are truly dangerous.

Can Antioxidants Help Clear Up "Smokers' Skin"?

Smokers usually have much of the rosiness of healthy skin drained from their faces—looking sallow if not parchment white. This is because the chemicals in smoke destroy the vitamin C and many other vitamins that have been ingested.

The body then goes after the lost nutrients wherever it can find them—and they are particularly plentiful in the tissues of the face. Drained of its nutrients, the face takes on the whiter color that make people look tired or a bit anemic.

How can smokers counteract the damage from the smoke if they do not wish to stop smoking?

A regular routine of antioxidant supplements will probably put a glow back in the face and help ward off some of the negative effects of cigarette smoke—but not over a very long period. Cigarette smoke, with all its toxic chemicals, is just too powerful a villain for any heroic healing supplement to fight forever and win, hands down.

Many more common inhalant allergies, such as allergies to ragweed, are caused by the body's production of *histamine,* a substance that fights supposed "invaders" like ragweed, that the body believes to be an enemy. The more histamine the body produces, the more of an allergic reaction the victim suffers.

Antihistamines help stop this allergic reaction but often produce many side effects, such as drowsiness and mental confusion.

Vitamin C, though, has been found to protect against histamine production. And all the antioxidants have been shown to suppress the inflammatory reaction caused by allergens. This is probably due to the fact that the antioxidants are so important to maintaining the health of the immune system. Apparently, the healthier the immune system, the less likely it is to become confused about what is a threat and what is not.

Caring for Allergies

Why should allergies be a cause of concern if they are just symptoms and not a life-threatening disease?

Because the inflamed, irritated conditions they can create in the tissues of the nasal passages, sinuses, and lungs can create breeding grounds for bacteria and viruses that may evolve into something more serious. If you take antioxidants for allergies, you may see the difference.

Antioxidants and Conquering Cancer

No one can say definitively that nutrient-based supplements of any kind cure cancer. But some of the evidence can't be ignored—and some of the case histories of cancer cure or remission are dramatic, and hopeful sounding. So while we wait to learn from the researchers whether antioxidants actually effect a cure, we can make some efforts in that direction. It's been proven that supplements can heal certain deficiencies that, if untreated, may *lead to* cancer.

What *is* cancer? Here's the simplest explanation of it:

Cancer is the result of errors in cell division

F.Y.I.

The unpleasant side effects of cancer treatments are hard on the healing body, but antioxidants can give weakened bodies a better fighting chance.

Chemotherapy and especially radiation therapy destroy vitamins A, C, E, and K. Supplements of vitamin E can prevent radiation burns and reduce scarring while vitamins C, E, and methionine can help the liver neutralize the by-products of destroyed malignant tissue.

caused by pollutants, free radicals, viruses and genetic predispositions to have those errors occur and duplicate themselves. The causes of cancer work much like video game villains; tearing through the body like bands of annihilators, disturbing tissues and cells to suit their needs.

The body's innate antioxidant forces defend against these invaders valiantly, but they can only do so much. When strengthened by antioxidant supplements, these fighting forces may win many of these battles and lower the cancer risk.

The list of pollutant-caused cancers is so vast it would fill many pages. For example: Residents of new mobile homes suffer increased rates of nasal passage cancer, probably because they are breathing the formaldehyde that is used in constructing these homes.

Residents of the New Orleans area suffer increased urinary tract cancers, traced, some experts think, to drinking water taken from the Mississippi River outside the city. Female residents of eastern Long Island suffer an unusually high rate of breast cancer and leukemia, probably because of air and water polluted by a nuclear waste facility.

It's necessary to add the qualifier *probably* to each statement because, while links appear to be very strong, there is absolutely no conclusive proof that these environmental conditions cause increased cancer rates in these geographic areas. But it may also be probable that the truth would cause economic havoc to businesses, governments, and people trying to improve their lot in life. New mobile homes, for instance, provide new-home luxury and clean, safe surroundings for many who could not afford such advantages otherwise, and people live where they do because of their jobs, family roots, and other considerations.

The notion of fighting cancer with nutrition has become an important step in our country's health care and is little understood. It's now believed that over 90 percent of all cancers are environmentally caused, and could be at least partially prevented by nutrition, much of it provided by healing supplements.

The antioxidants selenium, vitamins E, A, and C plus fiber (in food) have been shown in many studies to reduce cancer risk from environmental causes. As discussed in other chapters, people with deficiencies in the antioxidants were found to have a much lower incidence of cancer when these deficiencies were healed with antioxidant supplements.

This is not to say the other antioxidants don't work, but that the Big Four—selenium and vitamins C, A, and E—are most relevant.

The Basic Antioxidant Arsenal

Experts agree that these antioxidants should be taken *together* to work best as a fighting unit—that is, so to speak, a complex, more high-powered army of antioxidants. What's true is that the interaction between them tends to increase the individual potency of each antioxidant, whether vitamin or mineral. Or, as with selenium and vitamin E, one needs the presence of another to get the right chemical boost.

There's a common tendency to want too much too soon from antioxidants and seek the miracle cure—but hope should never prompt you to over-

Antioxidants and Cholesterol

Although controlling fat intake, reducing or quitting cigarette smoking, and refraining from overindulging in refined sugars helps *reduce* high cholesterol levels, so can certain nutrients and nonnutrient antioxidant supplements help prevent the formation and deposit of cholesterol in blood vessels.

Among these are the antioxidants taken in these quantities:

• Coenzyme Q10	60 milligrams daily
• Selenium	200 micrograms a day
• Vitamin E	200 to 1,000 IUs daily, with doses beginning at 200 IUs and slowly being increased
• Zinc	No quantity specified by experts
• Garlic	At least 2 raw cloves, three times daily; or 1 whole head roasted

indulge. Be smart and follow these limits so you don't overdo.

• **Selenium:** The maximum dosage is 200 micrograms. Amounts over this are toxic.

• **Vitamin A:** The maximum dose should be 25,000 IUs a day.

• **Vitamin E:** The maximum dose is 1,200 IUs a day.

• **Vitamin C:** Don't take more than 3,000 milligrams of vitamin C, spaced out into three separate doses, each of 1,000 milligrams. One dose of C can be taken with a multivitamin pill, one with an antioxidant complex, and one by itself.

Too much vitamin C can upset, and render useless, the power of A, E, and selenium.

Vitamin A

There are key vitamins and minerals that fight the oxidants produced in the body and they are:

Vitamin A, also covered in chapter 2—is called the "anti-infective" vitamin, since it is important to the body's ability to fight infection. It has a number of important functions in strengthening the immune system, for one, by enhancing the activity of white blood cells. Vitamin A also helps promote the healing of infected tissues and wounds.

The RDI of vitamin A is 5,000 IUs.

The Carotenoids: Beta-Carotene and Lycopene

These have been shown in many studies to help prevent cancer. Beta-carotene is often confused with vitamin A and is a "pre-formed" version of it. Beta-carotene battles "PUFA" oxygen radicals (that is, polyunsaturated fatty acid lipid peroxy radicals which can harm cells), and various forms of cancers. Beta-carotene aids in maintaining good health of the digestive tract linings as well as the lining of the vagina and uterus. Studies have also shown that women taking this supplement had reduced risks of heart attack, recovered better from angina, and developed improved cardiovascular health in general.

There is no Reference Daily Intake (RDI) amount listed for these carotenoids. For reasons of safety and efficiency, beta-carotene should be consumed only in food. Scientists have concluded that five three-ounce or larger servings of raw and/or briefly cooked yellow and red vegetables and/or fruit a day provide beneficial amounts of beta-carotene. Lycopene is found mostly in tomatoes, so two or three of those a day (washed carefully if not organic) will provide beneficial amounts of lycopene.

SMART DEFINITION

RDI (Reference Daily Intake)

According to the *Nutrition Desk Reference*, Reference Daily Intake—RDIs—is actually a new name for the old United States Recommended Daily Allowance (U.S. RDA). The name has been changed, but not the required amounts of the supplements.

Sarah, a thirty-eight-year-old New Yorker, has always suffered colds miserably. She read about the power of antioxidants in strengthening the immune system and helping to fight the common cold. When she came down with a cold, she went out and bought some supplements. Here was her regimen: After her breakfast she took 1 gram of vitamin C with flavonoids, 15,000 IUs of vitamin A, 400 IUs of E, 100 milligrams of selenium, and zinc lozenges, along with echinacea, an herb with antioxidant qualities. With her two other meals she took an additional gram of C.

Within three days, Sarah's cold symptoms were gone! She discontinued the zinc lozenges after three days and continues to take the vitamin C with flavonoids, the selenium, and the vitamins A and E.

Vitamin C

Vitamin C, also covered in chapter 2—protects against the damaging effects from the antioxidant superoxide and the hydroxyl radical and is considered a helpful fighter of respiratory infections and immune system booster. Because the body does not make vitamin C, it's in a constant state of requiring it from food or supplements to do its work.

The RDI of 60 milligrams can be increased safely to three grams a day, and twice that amount in the case of illness, according to Dr. Andrew Weil.

Vitamin E

This vitamin guards against "singlet PUFA" radicals (polyunsaturated fatty acid lipid peroxy radicals) which can harm cells. Vitamin E also helps repair damaged DNA and may prevent certain cancers caused by chemical toxins, cancer especially affecting the cervix, lungs, and gastrointestinal tract.

Vitamin E, to do its healing work, requires fat or oil in the digestive system, or it won't be absorbed. The body stores some vitamin E in the heart, testes, blood, adrenal and pituitary glands, and in fatty tissues. However, a diet too high in fat increases the risk of free radicals.

The RDI of 30 IUs can be increased safely to at least 400 IUs, according to experts. An overdose of vitamin E will cause sores around the mouth (especially in children), diarrhea, and gastric upsets. *Remember, all the dosages given in this book are for adults unless otherwise noted!*

Selenium

This mineral, best taken with vitamin E, is thought to help lower the problems associated with toxins and pollutants taken into the body, battles PUFAs, and is thought to help protect us from rectal, ovarian, cervical, and lung cancers.

Selenium helps maintain normal liver function and protein synthesis; it plays a role in male reproductive capacity (sperm count), and helps maintain healthy eyes, hair, and skin. In fact, there are prescription solutions or shampoos of selenium sulfide for the treatment of a common fungal infection.

Alcoholics may incur a minor selenium deficiency, as can those indulging too freely in refined and processed foods. A selenium deficiency is dangerous for pregnant women—it may result in miscarriage, infertility, and even possible post-delivery problems with discharging the placenta. But this group should not take extra selenium on their own without asking a doctor first!

The RDI of selenium is 70 micrograms. An overdose can be toxic.

Antioxidant Minerals

Zinc, copper, iron, and manganese also work against free radicals. As discussed in the chapter on minerals, they must be used prudently if you're thinking of them as healing supplements. More is not better: They can be toxic when taken in large doses. If these are included in a multimineral complex, their antioxidant protection is probably assured.

For dosage, check with your health-care practitioner. RDIs: zinc, 15 milligrams; copper, 2 milligrams; iron, 18 milligrams; manganese, 2 milligrams.

Antioxidant Enzymes

These help the body's regenerative processes and, in this case, are derived directly from food—that is, healing supplements are definitely part of a meal, too.

These important enzymes are specifically found in organic bean/seed/lentil sprouts. Broccoli sprouts alone are supposed to be twenty to fifty times more efficient than broccoli and other green vegetables in fighting cancer. These sprouts can be purchased at a good health food store, produce market, or even a better-quality supermarket. You can even grow your own. Be sure the sprouts are fresh and not limp or moldy before eating!

One ounce a week is said to be all that's needed of broccoli sprouts to get their health benefits. Two ounces of the other sprouts a week will provide their health benefits. Thus, an ounce of each in salads every day would be extremely healthful.

Glutathione

This is actually an amino acid—one of the building blocks of protein—but it's such an important antioxidant it's mentioned here as well as in chapter 6 on amino acids. Glutathione is made up of three amino acids whose job it is to mop up dangerous free radicals from fat-laden food.

One scientist claims glutathione destroys many cancer-causing substances. Vitamin C and selenium boost blood levels of glutathione, but this unique amino acid can be taken in single supplements.

There is no RDI listed for glutathione. Buy in tablet form at a reputable health food store, but be forewarned—it can be much more expensive than glutamine. It's safe to take 1 to 3 milligrams of glutathione a day, but ask your doctor first and follow the supplement label directions.

Glutamine

Glutamine is also an amino acid, and it contributes greatly to the strength of glutathione. The powder form dissolved in cold, nonacidic cold foods, like cereal or room-temperature rice, is the best way to eat it. Heat lowers the benefits of both glutamine and glutathione, so don't cook it. Never use glutamine if you are already ill.

There is no RDI listed for glutamine. About 5 grams of glutamine powder taken a day dissolved in any cold, nonacidic (acid and heat hurt glutamine) liquid is supposed to be the most inexpensive, safe, and beneficial source of glutamine, according to experts. Ask for it at your health food store. The best food source of glutamine, Brazil nuts is not at all a sure source if the nuts have been roasted or subjected to heat of any kind.

SMART SOURCES

Although glutathione can be found in most fruits and vegetables, the most abundant sources of it are in: *Acorn squash, asparagus, avocados, watermelon, and potatoes*

Coenzyme Q10

This is a newcomer to the healing-supplement rack, a fat-soluble antioxidant that's being looked at, in one camp, as one of the potentially great antioxidants—and in another camp by the dubious researchers who aren't sure about its healing powers.

Coenzyme Q10 is close to vitamin E in structure and is considered helpful in healing angina pain and oxidant damage to the body as a result of heart disease. It is also thought to help oxygenate the blood, thereby increasing mental clarity, and is an aid to circulation.

There is no RDI of coenzyme Q10 listed. Experts recommend a maximum dose of 300 milligrams—and no more—of this supplement a day. The supplement made in gel form is best absorbed.

Flavonoids

Flavonoids are chemical substances in certain plants, which gives yellow and orange fruits and vegetables their colors. There are more than five hundred kinds of flavonoids found in plant foods, almost all of which contain vitamin C.

Flavonoids help ward off colds and allergy and asthma attacks by destroying free radicals during the inflammatory stage of these conditions, and by using their antiviral and antibacterial powers to clear up the system.

There is no RDI listed. The best way to get enough of the flavonoids is to buy natural vitamin C containing them. You can also try these super sources of flavonoids:

Grapefruit Seed Extract

This is a new and reportedly exciting immune system booster that's packed with the bioflavonoid hesperidin. The extract, obviously from grapefruit seeds, can be taken in doses of 10 to 15 drops in water or juice up to three times a day.

No RDI is listed. Buy in pill form from a reputable health food store and take as directed.

Other supplements containing flavonoids that may help heal some common problems are:

Pycnogenol

This is a patented supplement distilled from pine bark, contains flavonoids, and is found effective in helping to relieve circulation problems, in clearing up or improving specific eye irritations and also gives the muscles more flexibility. Supplements come in dosages of 30 or 50 milligrams.

There is no RDI is listed. Buy in pill form from a reputable health food store and take as directed.

Tannins

Flavonoids are found in green tea, and according to some experts in herbal therapies, tannins are an important antioxidant and cancer fighter. Green tea, either taken as a hot drink or taken in tablets—you can buy green tea extract in this form—is especially good for protecting against radiation damage and in helping strengthen your immune system in general.

No RDI is listed. Green tea can be bought in capsule form and taken as directed. At least three cups of green tea a day are said to be necessary to provide the body with its health-improving effects.

SMART DEFINITION

Tannins
Compounds found in tea (particularly green and black varieties) and coffee, tannins bind iron and denature protein, in addition to their antioxidant qualities.

Many supplements used for other purposes are now being found to have substantial antioxidant qualities. In that sense they can be called "new" antioxidants even though most have been known and used for centuries.

Astralagus

A root used for healing in Asia, astralagus is especially popular in Japan and China. Astralagus is the root of a Chinese herb which raises the level and effectiveness of white blood cells—critical for immune system health. The root also prevents formation of the oxidant, lipid peroxide.

There is no RDI listed. Buy in tablet form at the health food store and take as directed.

Ginseng

A root that has been used for thousands of years by Asian naturopaths who have given it to their patients for a variety of problems, ginseng is most often for increasing energy, lowering blood pressure, improving circulation, and enhancing sexual function—especially for men.

Modern-day natural "menopause" formulas—intended to help establish female hormone balance—include ginseng in the mix. (See chapter 8 on healing women's problems for more information.)

There is no RDI listed. Buy ginseng in tablet, capsule, or liquid form at the health food store and take as directed.

Ginkgo Biloba

Ginkgo is one of the herbs thought to help circulation, but most significantly, it has been touted as an aid to memory and in maintaining mental efficiency. Ginkgo biloba has affected some interesting changes in the mental acuity of Alzheimer's patients—Alzheimer's being a tragic, so far incurable disease causing, primarily, loss of reasoning and general cognitive function.

Ginkgo biloba is a proven antioxidant similar in structure and content to the flavonoids.

There is no RDI for ginkgo biloba listed. Doses of 40 milligrams a day have been shown to be safe and effective.

THE BOTTOM LINE

Research shows that antioxidants are critical to health and help fight illness and the debilitating effects of aging.

Some antioxidants contain antiviral and antibacterial agents while other antioxidants protect against what may be a fatal illness—like cancer—or fight the inevitability of aging and the breakdown of tissues and bodily functions.

Before taking any antioxidant, whether it is vitamin A, E, selenium, or any other, get advice from a trusted practitioner in healing and nutrition.

CHAPTER 5

·····················

Amino Acids

THE KEYS

- Amino acids are the essential components of protein, and needed for cell replication and proper growth.

- Of the more than eighty aminos found in nature, only some of them are *essential;* that is, aminos that cannot be synthesized by the body and must be obtained in supplement form.

- The number one source of aminos is protein. You can get most of your amino acids from eating a balanced diet of meat and poultry, dairy products, vegetables, fruits, and grains.

- Amino acid supplementation used with weight training is a controversial topic, with much hearsay. Advice on training and aminos should come from an expert.

- Amino acid supplements, with the proper guidance, can be used for a wide variety of conditions.

Amino acids are critical for all forms of life—if not the very *basis* for what we're made of and what keeps us going: protein.

The human body, being an efficient and smart machine determined to thrive and survive, can utilize the range of amino acids found in all sources of food, no matter what the quality of protein—whether from animal meats, eggs, grains, beans, nuts, and vegetables.

In recent years, many of the aminos have been singled out from the pool and have been the subject of scientific scrutiny. This chapter looks at what you should know about the latest information on amino acids, not only to ensure good nutrition but as healing substances.

What Amino Acids Are and Why We Need Them

Simply, amino acids are the basic chemical units, or "building blocks," of protein molecules. Of the approximately eighty amino acids found in nature, the body needs only twenty of them for the hundreds of functions it performs for us every day, from memory retention to metabolism to growth and healing. To understand the chief function of amino acids, think of the process of renewal. You are you, but you're not made of the same cells that you were ten years ago.

Amino acids drive some of this cell replication. Cell renewal occurs because of instruction from your RNA and DNA, best described as the "software programs" that produce individual human characteristics. This is why we are each unique and not undifferentiated globs of living cells. Amazingly, DNA and RNA are synthesized—that is, produced—in the body with the aid of certain enzymes, which are, themselves, some of the many forms of proteins in the human body.

The process of cell replication happens at the microscopic level wherein cells replicate, or reproduce, any specific grouping. The DNA pattern that tells your body to grow new skin to heal a cut on your finger is different from the DNA pattern that tells your bones to produce marrow. If cells were not self-replicating, there could be no life that is either self-sustaining or able to grow—or age.

But how well people sustain themselves depends in part on finding and keeping a healthy amino acid balance.

Amino acids are found in two forms: the "L" form, as in L-carnitine or in the "D" form. The L form, derived from *levo*, the Latin word for "left handed," is *most useful* for the body. The D variety, derived from the Latin word *dextro,* or "right handed," means they are *not* as useful, except in certain cases. The amino acid forms referred to in this chapter are in the "L" form, unless otherwise noted.

In addition to their remarkable basic functions—sustaining life and growth—many amino acid nutrients have special healing properties. To affect healing, the amino acids that follow may be either taken alone, taken in combination with other amino acids, or with mainstream medicines.

F.Y.I.

Never self-prescribe any amino acid without first checking with your doctor, who knows your health history and can best advise you on what course of action to take.

Jill, a thirty-five-year-old boutique owner once skeptical of any nonpharmaceutical allergy "cure," changed her mind about treating certain allergies with tyrosine.

One weekend, Jill went for an extended hike with her hiking club, some of it through high grass. Her bare calves and ankles soon turned bright red—an allergic reaction to the grass.

When she returned back to the city that night, desperate for relief from the itchiness, Jill's friend recommended she shop for relief at a health food store—that is, she should try tyrosine.

Jill took two supplements of 20 milligrams and within twenty minutes her allergic reaction had cleared up. The itching stopped and the skin redness was fast fading.

The Amino Acid Family

There are about twenty amino acids needed by the body, and of them, nine are called *essential,* because they can't be made by the body and must be obtained through food or supplements.

These nine are:

Histidine, important for proper mental and physical growth and thought to aid in relieving symptoms of rheumatoid arthritis.

Isoleucine, important for growth in infants and nitrogen balance in adults.

Leucine, primary for growth.

Lysine, vital for growth and nitrogen balance and an aid in fighting genital herpes.

Methionine, required for sulfur production and normal metabolism and a serotonin stimulator said to help reverse sleeplessness.

Phenylalanine, needed for production of tyrosine, which is essential for growth and regulating nitrogen balance.

Threonine, vital for growth and nitrogen balance.

Tryptophan, needed for serotonin production in the brain.

Valine, important for growth and regulating nitrogen balance.

The other amino acids are called *nonessential* because the body *can* synthesize them. These amino acids and their basic functions include:

Alanine, important for proper liver function.

Arginine, important for kidney function, blood pressure regulation, and male sexual function.

Asparagine, important for brain functioning and nervous system health.

Aspartic acid, for assisting in the body's conversion of sugars to forms used more effectively.

Carnitine, for fat metabolism.

Choline, for neurotransmitter firing in the brain.

Cysteine, a natural sulfur antibiotic.

Glutamic acid, for healthy brain function.

Glycine, for brain function and relief of gastric acidity.

Proline, to create collagen and provide absorption of nutrients in the intestines.

"Mind" Aminos

While the pharmaceutical industry is working on producing a whole new generation of "smart" drugs called nootropics, which means they act on the mind, three amino acids—phenylalanine, choline, and tyrosine—are proving to be strong contenders in the race toward increasing brain power.

These amino acids may improve concentration, problem-solving abilities, and counteract effects of excessive narcotic, substance-abuse drugs (such as cocaine). Cocaine and amphetamines, for example, deplete the supply of certain neurotransmitters in the brain, but they may be restored by taking amino acids phenylalanine and tyrosine.

They're also being examined for their abilities to cure and/or prevent Alzheimer's disease and the effects of aging on the brain. Phenylalanine and choline, especially, may help mental clarity since they are converted into the neurotransmitters norepinephrine and acetylcholine, critical carriers of messages between brain cells.

Serine, for digestion, blood coagulation, immune reactions, and fertility.

Taurine, for promoting necessary chemical reactions in the body.

Tyrosine, for skin, thyroid, and production of hormones that deal with stress.

Sources of Amino Acids

Think steak, eggs, avocados, peanuts, whole wheat rolls, and pumpkin seeds—even bagels and cheese. While all these foods contain amino acids, you need to know which may be best for you. Most people get a sufficient supply of their aminos from eating a balanced diet that includes, meats, poultry, and fish; dairy products; vegetables; fruits; and grains. But not all people in our fast-paced, complex age have the opportunity to ensure that their diets are meeting their nutritional needs. For some, particularly vegetarians, amino acid supplementation may be the answer.

The Number One Source: Protein

There's no way around it: *The* best source of all nine essential amino acids is in *high*-quality protein or complete protein sources from animal products, that is, meat, eggs, and milk. The proteins from animal products can, and have throughout history, pretty much sustained human life—never mind a balanced diet, which includes fruits and grains.

Indeed, many groups of people have always existed on nothing but animal foods. For centuries, both certain Eskimo groups and some northernmost Native American tribes seldom saw, let alone cultivated or ate, any but very few fruits or vegetables.

Seal liver may have been high in cholesterol, but so did it provide the vitamin C rarely found in plant food in the Arctic climate. And holding to tradition in another part of the world even today, the mainstay of the African Masai diet is a drink of blended cow's blood, milk, and butter. All are high protein and rich in amino acids.

Other than soybeans and a few grains, no plant food contains all the essential aminos.

At the next stage, there are nuts, grains, and legumes—that is, beans, peas, lentils, and peanuts—which provide what are known as *medium*-quality proteins. This means that the amino acids found in these foods can sustain life but are not complete enough to promote proper growth. There is one exception: Among all the nonanimal sources of protein, *soybean products* come closest to being a high-quality protein.

Finally there are fruits, vegetables, and gelatins, which are *low*-quality proteins. This means they can sustain life, but not for long. Vegans, those vegetarians who consume no animal products at all, must get their high-quality proteins by combining certain types and quantities of plant proteins that, together, become complete proteins of the type necessary for sustaining life.

Vegans need to research the amino acid, vita-

F.Y.I.

The good news for vegetarians is that the body maintains an amino acid pool, so that you can combine incomplete grain, vegetable, and dairy proteins and come up healthy. For example, if you eat brown rice and vegetables at one meal and beans or lentils later in the day, you've ensured normal protein metabolism. All essential amino acids are found in a combination of beans, grains, and vegetables. Ask your nutritionist to be sure, however.

How Much Protein Is Right for You?

Amounts of protein needed by the body can change with certain conditions. Always consult a doctor about both requirements for and adjustments in protein intake for children, for women during pregnancy and lactation, and for men and women during times of illnesses, stress, and periods of dieting and weight loss.

Most medical sources agree that 60 grams of mostly high-quality protein a day is both sufficient and beneficial for adult women of normal health, while 80 grams are recommended for males. Too little protein can affect changes in health and appearance.

Is a low-protein diet ever a wise choice in terms of getting the amino acids you need? One massage therapist revealed in a magazine article that she believes the muscle tone of women and men has deteriorated over the past ten years. Flabby tissue, she says, now replaces the firmer bodies people used to have. She attributes this change to the low-protein, high-carbohydrate diets that came into vogue in the 1970s and still continue to influence certain dieters.

Warning! Excess protein intake, especially from high-fat, high-cholesterol animal sources, has been associated with a number of diseases, such as atherosclerosis (narrowing of veins and capillaries from plaque buildup) and certain cancers. Never eat protein foods above the recommended daily allowance in attempts to get a greater supply of amino acids.

min, and mineral content of the foods they eat or else they may bring on nutrient-related conditions such as confused mental states and low energy.

Best Sources in Supplements

Just as vitamins and minerals are found in certain balances within the body, so is it with amino acids. You can buy balanced amino acid concentrates most commonly in liquid, powder, or tablet form. There are formulas that contain all the essential and nonessential aminos—which takes the worry out of getting all the nutrients at once—as well as individual amino acids or those made in combination formulas.

The powders may come from animal sources and include such low-fat ingredients as dried skim milk and egg whites; or from plant sources, usually with soy powder as the main ingredient.

Before you buy protein powders, always check the amount of carbohydrates in them. Any carbohydrate level over 3 grams per ounce usually means the powder is loaded with sugars and other fillers that you do not need. In fact, such formulas can be harmful, especially for those with diabetes or hypoglycemia, or for those who are obese.

Amino Acids and Weight Training

In the quest for bigger, better, and more "ripped" bodybuilder looks, both men *and* women who aspire to this extreme muscular state often believe that taking more protein and amino acid supplements will help them achieve this goal faster and better.

This belief is dangerous. For one, cancer has been associated with high protein levels in the body—and better bodies do not necessarily spring from mega-amino and protein gluts. Protein intake in bodybuilding is not a matter of quantity but of quality, frequency, and kind of amino acids taken.

In an article in *Muscle and Fitness Magazine,* Dr. Barry Finnin and Samuel Peters reported that the

SMART MONEY

Should parents give their growing children amino acid supplements or not?

Jennifer Henkind Ferraro, M.D., a board-certified pediatrician in private practice in Stamford, Connecticut, says, "*No.*" Amino acid balance is especially critical in growing children and they can get what they need through a well-balanced diet. There's not enough research on how individual amino acids affect a child's developing body and parents *should not experiment* with these supplements.

"A balanced diet is best," the doctor says. "And if a child's diet is lacking in one of the basic food groups—like protein or fruits and vegetables—then a supplement can be beneficial."

kinds of amino acids bodybuilders take to reach their physique goals and muscular strength are an entirely different class from those taken by average people. These are more quickly absorbed and are available in powdered supplements that cost a great deal of money.

The special quality of these amino acids, according to Finnin and Peters, is "bioavailability." This means that these amino acids are processed for rapid, efficient delivery to the muscles without having to be digested first—or what is known as "free form" aminos. Protein drinks offer a similar bioavailability in that their protein is *hydrolyzed,* which means predigested.

A normal meal of protein-packed foods does not get the protein into the system until a few hours after eating. This is because the liver, which is the protein-processing organ, can handle only a few grams of protein at a time.

The Amino Acid Supplements

Can amino acids be taken individually in supplement form? Here's what experts say on the subject:

A powder or liquid containing a *balanced complex* of amino acids would be most experts' recommended supplement of choice. Here, though, are profiles of nine amino acids that either have been in the news, are the subject of debate—especially as to their multifunctional properties—or are of special interest to most people. These aminos are found in either individual or combined formulas.

Remember: Always speak to your health care

professional before taking any amino acid separately or in a complete formula for special health considerations.

Arginine

Arginine, which converts into the gas *nitric oxide* in the body, has for years been used in hospitals to save "blue babies"—babies who suffer from impaired breathing because the airways of their lungs have not opened properly.

This conversion from arginine to nitric oxide is proving to have a number of beneficial effects beyond those associated with saving infants. The benefits include the ability to help maintain normal blood pressure levels and perhaps reduce high blood pressure and its attendant risks—such as stroke—as well as helping to keep coronary arteries open, maximizing heart health in both men and women.

For men, arginine supplements have been shown to help increase blood flow to the penis, thereby—for some men—enhancing erections and, if they are so afflicted, relieving some impotence problems. Arginine is also being researched in terms of boosting sperm count and even reversing male infertility.

One dose of a 500-milligram tablet a day is considered safe. Speak to your doctor before increasing your intake for any reason.

STREET SMARTS

A newly converted "vegan," Jane found herself suffering incredibly painful bouts of weakness and depression after four months on her regimen. She spent weekends on her couch, enervated and sobbing.

A concerned friend took her to brunch and ordered an omelette for both of them and a glass of milk for Jane. Too weak to protest, Jane began eating; halfway through, she felt herself beginning to recover.

So restored by these 30-odd grams of complete protein— eggs contain 6 apiece, milk about 8 grams per ounce—that she was able to laugh for the first time in weeks. She has since added at least 40 grams of complete protein a day to her diet, mainly from egg whites and protein powder from animal sources.

The Occasional Amino Acid "Snack"

Amino acids also come in powder form and can be taken as suggested on the label, or, they can be whipped into a "pudding" with the addition of water and honey. For some, this mixture makes for an *occasional* meal substitute, or snack.

There is no easier way to get a quick and complete protein meal but remember: These amino acid products should never be regarded as a replacements for meat—they lack iron and other nutrients unique to animal protein.

Don't overdo on amino acid powder! Too much protein—more than about 80 grams a day—may be a cause of serious health problems.

Carnitine

If sluggish fat metabolism is a problem, you may want to look into carnitine's properties. Known for metabolizing fatty deposits, some research shows that carnitine is especially effective for those with obesity problems. It is also helpful in reducing the health risks that may occur in diabetics who do not metabolize fat properly.

In a Japanese study, carnitine was found to maintain blood glucose levels in subjects who were in the extreme of dieting—fasting. Since plummeting blood sugar levels are what often make dieters so irritable, carnitine's ability to regulate blood sugar makes it a possible mood enhancer for people who are dieting. Carnitine is also said to improve brain functioning.

Carnitine, for most purposes, can be taken in amounts of 200 milligrams three times daily, and in amounts of 400 milligrams three times daily for improvement in brain function, fat-deposit metabolizing, and other benefits mentioned here.

Warning! Although up to 3,000 milligrams of carnitine a day is considered nontoxic, it should *not* be taken with herbs like ginkgo or guarana, which are stimulants. Combining carnitine with stimulants can raise blood pressure to dangerous levels.

Cysteine

This antioxidant's qualities enable it to protect the body against mercury buildup. (We can inadvertently ingest mercury by eating foods contaminated by it during the growing process. Many dentists still use mercury compounds to fill teeth.)

Cysteine is also reportedly the best antidote for alleviating the effects of an acetaminophen overdose. An ingredient found in such products as Tylenol and other nonaspirin analgesics, acetaminophen creates large amounts of free radicals in the body when taken in doses greater than those recommended by a physician or noted in the directions on the bottle's label.

Check with your health-care practitioner for dosage guidelines.

Glutamine

Glutamine is especially important in detoxifying oxidized fat and in keeping the lungs, heart, liver, and blood cells from being damaged by free radicals. It is used in supplements in Japan to ward off effects of environmental and other forms of pollution that cause free radicals.

Glutamine is also one of the contenders in the

F.Y.I.

Cysteine, glutathione, and taurine: This triumvirate of amino acids can together perform antioxidant functions to help heal the damage caused by free radicals in the body, some of which can lead to cancer.

battle to defeat the inevitability of certain aging factors. That is, it is considered an anti-aging supplement. Although all the amino acids are needed by aging people, glutamine is proving to do the most for age-related problems.

Besides strengthening the immune system by blocking damage to the cells, this amino acid breaks up unmetabolized fat. Glutamine also is said to ward off serious eye problems of the elderly, such as macular degeneration.

Glutamine and tyrosine together also may alleviate depression, but these take longer to work than does phenylalanine (see below) and are better compared to such antidepressants as Prozac in their effect.

Vitamin C in amounts of at least 100 milligrams a day and selenium are said to boost the level of glutathione in the body and may thus boost the effect of the glutamine supplement.

Glutamine can be taken in amounts of 1 to 3 milligrams daily, which does more to raise glutathione levels in the body than taking glutathione directly.

Methionine

This is especially good for helping to remove heavy metals, such as mercury, from the body or to help selenium absorption and help prevent atherosclerosis—the buildup of cholesterol plaques in the arteries and veins, which unchecked may lead to heart attack or stroke.

Osteopath Dr. Leon Chaitow suggests doses in a range of from 50 milligrams to 500 milligrams a day, but the higher doses must be used with caution and only under a doctor's care. Methionine

Homocysteine and Heart Disease

There is one way an amino acid can contribute to heart disease: A metabolite of methionine, homocysteine, causes cells to decrease their production of clot-preventing and clot-dissolving substances and increase production of clot-promoting substances.

The presence of homocysteine in the body has come to be a predictor of heart attack, stroke, deep vein thrombosis, and other circulatory problems that are very serious health risks. It would be wise to get your blood tested for homocysteine levels.

The presence of this unwanted by-product of an amino acid is caused by a deficiency of B vitamins, especially vitamins B_{12}, B_6, and folic acid. The homocysteine level can be lowered by taking all three vitamins properly balanced in a good B-complex or multivitamin supplement. Making sure to eat foods rich in vitamin B, such as liver, dark green vegetables, and legumes, will also help enormously.

Note: Some physicians may try to discourage the testing of homocysteine levels in the blood, but other medical experts *insist* upon it and call it life saving. Decide with your health-care practitioner which is right for you.

should be taken with vitamin B_6 to avoid having any extra methionine cause a buildup of homocysteine.

Phenylalanine

The reason for food cravings—that intense desire for a particular food—can have many causes. Cravings can arise from sugar addiction or even a food allergy and have nothing to do with the body's need for nourishment.

One way to curb food cravings and the urge to overindulge is with the amino acid phenylalanine.

The Creatine Factor and Bodybuilding

What *is* creatine phosphate and what does it do?

The supplement creatine is made from three amino acids: arginine, methionine, and glycine. Together they work to keep the level of creatine phosphate high in the bodybuilder.

Creatine is a nitrogen-containing compound that combines with phosphate to burn a high-energy compound stored in the muscle, thus giving more energy in training.

A bodybuilder needs constant supplies of "CP," as creatine phosphate is referred to, in order to replenish the body's stores of adenosine triphosphate (ATP). ATP is crucial to the energy maintenance of all human cells. The body's stores of ATP are depleted every time a muscle is contracted.

Creatine should never be taken without a doctor's approval—whether or not you want to bodybuild. Anyone embarking on a bodybuilding regime is advised to consult a doctor, particularly one trained in sports medicine, as well as a certified trainer.

The amino acid program required is complex, expensive, and needs careful qualified supervision to work. Ask a physician for referrals to a good facility or trainer.

Many over-the-counter diet drugs contain L-phenylalanine, thus reversing this very unpleasant condition immediately.

But because of its tendency to raise blood pressure as it suppresses the appetite, phenylalanine should not be taken more often than directed or ingested in greater doses. Its appetite-suppressing effect wears off after a week or so. It is best to avoid the phenylalanine-based appetite suppressants that also contain caffeine—these might be especially likely to cause high blood pressure.

Also useful in curbing food cravings is a powder supplement containing phenylalanine, along with all the amino acids. An ounce or two taken every two hours may help keep zigzagging blood sugar levels—another cause of food cravings—in check.

So can this amino acid prevent the "feeling of starvation" syndrome that stimulates hunger and makes the body slow the metabolism (burning of calories) to a halt. Low-calorie fruits and vegetables can be taken with the powder at mealtime to make the combination more balanced, nutritious, and flavorful.

Addiction-prone people will find a protein-replacement regimen supplemented with amino acids to be helpful without any jolts or drama at all. Hopefully, they will eventually enjoy its benefits in helping to create a balance in the body and lose their cravings for such drugs.

Another form, *D*-phenylalanine, prevents breakdown of endorphins—the "feel good" substances that prevent pain—in the body. Phenylalanine in D- and also in L-forms, found as "DLPA," or D-phenylalanine may mitigate the pain that arises from any number of disorders or diseases.

Another of phenylalanine's supposed benefits is its ability to alleviate depression. Phenylalanine has often has an effect on depression that is fairly immediate, like that of amphetamines or certain narcotics.

Phenylalanine should not be taken by anyone who has high blood pressure or who is taking monoamine oxidase inhibitor (MAO) drugs. These are substances that are found in prescription antidepressants and certain foods and wines.

This amino's standard dosage for producing the appetite-suppressant effect is 2 tablets of 475 milligrams of phenylalanine before each of three meals a day. Phenylalanine in amounts of 100 to

SMART MONEY

"Phenylalanine's appetite-suppressing effects are remarkable," says osteopath Leon Chaitow, D.O., N.D. But besides this effect, Dr. Chaitow adds, "Phenylalanine is also known for increasing positive feelings not usually associated with dieting, such as alertness, sexual interest, enhanced memory, and after twenty-four to forty-eight hours, an antidepressant effect."

Fighting Addictions to Stimulants: Can Amino Acids Help You Kick the Habit?

Caffeine, amphetamines, and cocaine, among other stimulants, work in at least two ways to create their effects in the body. They act as catalysts to release the brain's own natural stimulants, and they temporarily raise blood sugar in a way that provides an instant jolt.

Addiction occurs when larger and larger amounts of the stimulants are required to produce the desired "high" or "jolt."

Can amino acids help break the habits?

Research shows that the amino acids phenylalanine and tyrosine work to reduce stimulant craving by releasing the brain's natural stimulants on a more consistent basis.

Here's the smartest way to take these amino acids to produce this more natural effect: Instead of taking the stimulant, substitute for it an ounce or two of protein powder mixed with water, no more than three times a day. Try this formula for at least four days, until the caffeine or other stimulant is detoxified out of the body.

Warning! Many stimulant addicts are hooked by the sudden jolt of the stimulant as much as they are to the feeling of being energized. People who use amphetamines, for example, often miss the sudden high produced by them, and so turn to caffeine as a substitute—also because of the instant "buzz" it produces.

500 milligrams daily—*but no more than this range of dosage*—for up to two weeks, *but no longer,* can be taken for appetite suppression and depression.

Taurine

Although a less potent antioxidant than glutamine, taurine is still important for its free-radical-fighting properties. Taurine has been found to be an important destroyer of a free radical called hypochlorite, which is produced by and becomes prevalent with autoimmune diseases and infections. More research is being done on this amino acid to explore its defense properties.

Tryptophan

Tryptophan is famous for its ability to reverse sleeplessness, thereby allowing the body to heal the effects of sleep deprivation. To be effective, however, tryptophan must be taken alone—that is, neither as part of an amino acid powder nor with foods containing protein. When it's taken alone, it doesn't have to compete with other amino acids to get absorbed into the system.

This amino acid works by producing serotonin, the neurotransmitter said to cause all kinds of desirable states, from relaxation to euphoria.

There's a catch here: Because a contaminated batch of tryptophan sold in 1989 was found to be fatal, tryptophan, not the careless manufacturer, was blamed. Tryptophan was taken off the market and is not available in a single over-the-counter supplement tab.

Interestingly, the same serotonin-synthesizing effects produced by tryptophan may be produced by a combination of vitamins B_3, B_6, and C, particularly B_6. Never take more than 200 milligrams of B_6 a day for longer than six months because more prolonged ingestion of B_6 can impair the nervous system. Dosages of 1.5 to 3 grams of tryptophan are best for insomnia; any higher a dose may render the tryptophan ineffective.

F.Y.I.

Never give your child tryptophan if he or she has trouble sleeping. If your child has a sleeping disorder, says Connecticut pediatrician Dr. Jennifer Henkind Ferraro, the reason is probably due to something other than a natural deficiency in tryptophan—and "you need to look down other avenues" for causes and solutions. "It's much safer to give your child a small cup of chicken soup rather than tryptophan to relax him at bedtime."

Tyrosine

Tyrosine is a powerful histamine blocker and a good source of allergy relief that has almost no side effects unless taken with MAO inhibitors. And like most supplements, tyrosine is *not* for everyone: It is especially to be avoided by anyone with asthma. Most asthma sufferers need to be careful of antihistamines and histamine blockers.

According to Dr. Leon Chaitow, tyrosine has been found useful in treating Parkinson's disease and depression. While some sources warn against taking tyrosine with the natural mood elevator St. John's Wort, it is nonetheless one of the ingredients found in some St. John's Wort preparations, such as Natrol's Mood Support. A chemist employed by Natrol insisted tyrosine is safe in combination with—and enhances the effect of—St. John's Wort. Ask your physician or nutritionist to be sure.

THE BOTTOM LINE

Amino acids in a proper balance are crucial for life, and each of the twenty the human body utilizes has unique properties. If you opt to take them individually or in specific combinations, you must consult a nutritional expert who is knowledgeable of the latest amino acid research to set up a supplement program for you.

If you are advised to take amino acid supplements, the best way is in complex form—that is, in capsules or powder made from eggs, milk, and other animal products, which provide the most nourishing kinds of proteins. The soybean is the only complete source of protein in the plant family.

The Healing Plant Kingdom

H erbs, roots, fruits, or algae—they're all classified as both herbs *and* plants. You might think that many of them belong in a book on food and nutrition.

But only some these herbs can be eaten; others are medicinal, while yet others are literally garden-variety decorative and food only for the bees. Some can be used to make healing supplements. There are so many healing herbs that a description of all would take several volumes. This chapter presents the most common of the healing herbs and the conditions they can help.

Natural Healing from the Plant Kingdom

The healing properties of plants come from the chemicals produced from their various parts—the stems, roots, leaves, or flowers. The knowledge about how these plants help heal has come down through history and from many world cultures.

Herbals can be used in many different forms: Boiled in teas, either in tea bags or loose plant parts; ground up in powders and sprinkled on food; dissolved into liquid tinctures; and rubbed onto the skin in the form of salves and ointments. Every good health food store has rows of herbs in apothecary-like jars, carefully labeled. These are the original herbs, most of which must usually be processed in some way before they can be used. It's definitely more convenient, and probably more sanitary, to buy herbs in commercially prepared forms, such as tea bags, tablets, or tinctures.

But what do you need to know when buying herbs? Here's a simple-to-follow, important guide.

Shopping for Herbs

Since herbs are not regulated by the FDA, they should be purchased from as reputable a vendor and manufacturer as possible. Beginners are urged to buy herbs in tablet or capsule form that contain few fillers. Ask your health-food-store source for advice if all else fails. Dr. Alex Cadoux, a medical doctor recognized for his wide knowledge of alternative medicine, warns that herbs can be ineffective, if not dangerous, when used in raw plant form.

The herb supplement container should always contain the word "standardized" on the label. That means the herb used has been processed in a way that provides a standard amount in all capsules of at least one of the major herb ingredients, such as the ingredient hypericum in St. John's Wort. Yes, these standardized supplements are more expensive, but well worth it. Even standardized herb pills should be taken immediately because they supposedly lose potency with age.

Smart Tips for Buying Herbs

The following products may or may not contain herbs that have been standardized. Be sure to check the label or ask the seller.

• **Tinctures of herbs** (sold in liquid form with droppers) usually contain a high concentration of

F.Y.I.

According to Alex Cadoux, M.D., noted alternative medicine authority, about 50 percent of all pharmaceuticals are derived from common herbs.

A Crash Course in Herbology

The use of herbs and plants for healing modalities is as old as the first human who chewed on a leaf and got a response he did or didn't want. Eventually, he learned how to cultivate that which he needed for his and his tribe's benefit and that which he grew to use to gain power.

Anthropologists also say that humanity first discovered the healing properties in plants most likely through observation and, riskier, trial and error. Early humans probably watched their indigenous animals, grazing or plucking berries and nuts, and knew what was safe. In parts of the world, man might have actually observed primates using certain herbs to treat skin parasites or saw cats gravitating to growing catnip and watched the response of the animal who immediately rolled on its back in some sort of feline high.

After learning which plant did what, and more important, what *healed* what, certain specialists arose in tribes and became healers or herbalists, known for their skill, instinct, and ability to both diagnose illness and heal with plants.

These herbalists learned to prepare the herbs in various ways, categorize them according to effect, and preserve them by drying. Later, herbalists began recording the effects of various dosages of the herbs and passed all this information on to apprentices.

In America, the herbal arts have been sneered at by most medical doctors and, until recently, they were nearly forgotten. Then less than ten years ago, the medical profession was forced to face a few facts: Herbs were not only still used in Europe, but widely researched and more respected abroad than ever. Germany even has an official government agency for research on healing properties of herbs. Medical doctors in America finally had to face the fact that more and more people were taking herbs on their own.

Increasingly worried about the use of prescription pharmaceuticals, including ever stronger antibiotics, and of unwanted side effects and medical costs, more and more people consider herbs as drug alternatives.

Now that Drs. Andrew Weil and Deepak Chopra, among others, have made alternative medicine more mainstream, other medical doctors have become more open minded. Echinacea for its strengthening effect on the immune system and St. John's Wort for its antidepressant effects are probably the herbs most commonly recommended by traditional M.D.'s. at this time.

alcohol (around 40 to 50 percent) and are very expensive. A nonalcoholic liquid used in tinctures, glycerin, is said to make the effect of herbs far less powerful than does alcohol.

• **Herb teas** in prepackaged form are also a good source of dried herbs. The loose teas are usually less expensive than the tea bags, but the bag variety assures proper quantity. Also, the used tea bags often have uses in external application.

• **Herbs in bulk or sold by weight** are available for those who don't want to buy prepackaged supplements, and can be found in such shops as botanicas and health food stores. They can be consumed in various forms.

The more experienced herb connoisseur can buy herbs in bulk form and prepare her own teas, tinctures, or capsules, or just take the herbs straight sprinkled over food or in juice. A health food grocery store, the kind that sells fresh organic foods, is probably the best place to find the freshest quality herbs in bulk. Empty capsules, tincture dropper bottles, and tea bags are usually for sale in a health food store. The herb shelves usually hold jar after jar of fairly clean, loose herbs, carefully labeled.

Taking Herbs

Unlike vitamins, minerals, and amino acids, which are best taken in balanced complexes, herbs are usually best if taken separately. Multiherb formulas often contain so little quantity of the various

F.Y.I.

As they would after taking any new medication, those taking herbs should be alert to any allergic reactions to them. Skin rash, difficulty swallowing, and especially difficulty breathing are cause for calling an ambulance because these symptoms could mean an allergic reaction so severe it could be fatal. Happily, people rarely have such reactions, but the allergy symptoms should be watched for nevertheless.

herbs that they cannot provide enough of any one to be of real benefit.

Follow the herb bottle or package label directions to the letter:

• If a label advocates taking one pill three times a day, that does not mean taking three pills once a day, it means spreading out the dosage over a day-long period so that the body will get a more prolonged benefit.

• Do not take more pills than the directions suggest. More is not only ineffective, but often dangerous. On the other hand, do not take fewer pills—say, one every few days when you remember to take one—and expect any substantial results.

• Do not be impatient. Most herbs are unable to provide overnight magical cures of the sort the American medical patient has come to expect from potent antibiotics, painkillers, allergy relievers, and other instant-effect drugs.

• The herb user must adhere to the suggested daily dosage schedule, even if it seems there will never be any results. The benefits will usually occur and almost always be worth the patience required.

How to Prepare Herbal Teas

Here are some definitions of methods that have been used for thousands of years to prepare herbs for consumption that have been prepared in boiling water:

• **Teas:** To make a tea, an ounce of dried herb is dropped into a pint of boiled, hot water for a certain length of time, usually several minutes, then the tea is strained and drunk.

• **Infusions:** To make an infusion, an ounce of dried herbs is dropped into a pint of boiled water, then steeped for at least fifteen minutes, usually more, before the tea is strained. The longer the herb is left in the water, the more of its phytochemicals—that is, naturally occurring active plant nutrients—will infuse into the water. Teas and infusions are best for herbs whose flowers and leaves contain the phytochemicals.

• **Decoctions:** To make a decoction, an ounce of dried herbs is dropped into a pint of water while it is boiling. After, it is left to simmer for at least a half hour. Decoction is used for herb phytochemicals found in the roots and bark of plants.

Herbs That Enhance Mood

• **St. John's Wort** is a popular mood enhancer, sedative, and alterative. Like Paxil, Prozac, and other prescription antidepressants, St. John's Wort has antidepressant effects that sometimes take two to six weeks to show results. Most doctors recommend taking one pill of 300 milligrams containing 0.3 percent hypericin—considered to be the active chemical in this herb. According to Dr. Steven Brantman, M.D., an authority on this herb, it was thought that St. John's Wort worked by

SMART MONEY

"Parents should *never* concoct herbal brews of what they think their children need, because it can be dangerous," says Connecticut pediatrician Jennifer Henkind Ferraro, M.D. "Avoid herbal brews that can cause traumatic reactions."

inhibiting the monoamine oxidase (MAO) enzyme, but research showed its MAO-inhibiting effects were simply not strong enough.

Another theory is that hypericin produced feelings of well-being by inhibiting serotonin reuptake in the cells—this means it allows serotonin to take its rightful role as transmitter of "good feeling" messages to receptors in the brain. (Serotonin is the neurotransmitter amine produced from the amino acid tryptophan that leads to a calming state). However, further research revealed that St. John's Wort produced this effect even when the hypericin was extracted. "Thus," says Dr. Brantman in his book *Beat Depression with St. John's Wort*, "the only safe conclusion is that we really do not know how St. John's Wort works in treating depression."

• **Cloves** taken in any form can produce a mild sense of euphoria.

• **Licorice root** has compounds that are instant mood lifters. However, too much licorice can also can lead to hypertension. Never drink more than three cups of licorice tea a day or overeat licorice candies.

Note: This herb can, especially in amounts over two or three cups a day, cause headaches, depression, sodium retention, water retention, potassium loss, and high blood pressure.

• **Passion flower** can elevate mood when taken as a tincture in warm water four times a day. Dr. Andrew Weil recommends one dropperful of passion flower prepared this way.

• **Rosemary,** as a spice in cooking, may increase

the happiness of dinner guests—and is also considered a digestive aid and antioxidant. Never ingest rosemary oil, although a few drops of the oil in a hot bath is a relaxant.

The Phyto Fighters

These herbs are in a separate section because they double as foods. Cereal grasses are complete foods in concentrated form. They contain all the health-giving properties of the dark green vegetables we need.

Three grasses are particularly popular for human health: barley grass, wheat grass (including kamut), and alfalfa. You can buy these grasses fresh in good health food stores or in juice or supplement form. Do not economize on price.

• **Alfalfa** is one of the richest mineral foods known. It has abundant amounts of nutrients needed by the human body. It helps to sustain well-being and to treat illness. It is best known as a base for liquid chlorophyll.

• **Barley grass** is an excellent source of the broad spectrum of nutrients for the growth, repair, and well-being of the body. One of the remarkable healing benefits is DNA repair.

• **Wheat grass juice** has been reported as being the richest nutritional liquid known, containing the greatest variety of minerals, trace minerals, and vitamins of any vegetable. Some people suffering from rheumatism claim to have been cured by wheat grass therapy.

F.Y.I.

All herbs can be dangerous to pregnant women and to their unborn babies, and to lactating mothers and their nursing babies. No herbs should be taken during these times without a physician's approval.

Properties of Herbs

Herbalists describe the properties of herbs in a number of classifications. These are some of them. This helps you, too, know more clearly what effect the herb has:

Analgesics relieve pain and some may also have muscle-relaxant qualities.

Alteratives are used as blood purifiers, to rid of toxicity to the system.

Carminatives relieve gas and griping.

Antacids neutralize "acid stomach."

Diuretics increase urine flow.

Nervines calm nervous tension.

Expectorants help expel mucus from the lungs and throat.

Sedatives quiet the nervous system.

Tonics promote the function of body systems.

Algae: Fighters for Good Health

The amazing algae is a whole food containing all the nutrition needed to sustain life, replete with protein, carbohydrate, essential fatty acids, vitamins, minerals, antioxidants, and other phytonutrients, including chlorophyll. Merely adding 1 to 3 grams of algae to regular food intake each day can improve quality of life.

Algae has therapeutic as well as nutritional value. Published data indicates that it lowers cholesterol, enhances liver function, improves immunity, lowers blood pressure levels, *increases* beneficial bacteria in the intestines, normalizes glucose levels, and may even prevent or lessen the severity of cancerous tumors. Algae contains a plethora of vitamins and minerals too numerous to list here. Here are two especially valuable forms of algae that can be taken in supplement form:

• **Spirulina** is a blue-green algae that is one of the oldest plants on the planet. It is both a plant and a bacteria. Products containing spirulina have been growing in popularity for nearly twenty years and have been bringing health benefits to people in over forty countries.

• **Chlorella** offers the unique property of fighting constipation by stimulating peristalsis and encouraging the growth of beneficial bacteria. It detoxifies the intestines of pollutants, including heavy metals, by binding with them and then eliminating them from the body. The cell wall material of chlorella stimulates production of interferon in humans, the body's first line of defense.

A Directory of Herbs and the Common Conditions They Can Help

• **Aloe** plant leaf pulp helps heal and soothe bites and burns. The dangers of contact with poison ivy, oak, and sumac can be mitigated with the application of pulp from a slit fresh aloe plant leaf. It is only the medicinal aloe plant that has healing properties (it has straight, narrow green leaves), not the decorative aloe (with the curling side leaves).

The medicinal aloe plant is also effective in healing burns. Cut off a low leaf from a medicinal aloe plant, spoon out the pulp from the lowest part of the leaf, and apply it to the burn. For sunburn, compresses of raw aloe pulp applied to clean sunburned skin every day will help decrease the pain, redness, and chance of peeling.

Aloe vera also comes in juice or gel form: When taken internally, the mucilaginous pulp of the aloe vera acts as a laxative, cleans out the colon, clears the throat and is said to lessen cravings for sugar and caffeine by helping regulate blood sugar levels. Look for products that have 98

F.Y.I.

Buy algae and grasses from a reputable health food store. The best brands cost more but are worth the price.

Flower Remedies to the Depression Rescue

Every health food store has a selection of what are called flower remedies, usually in tincture form, for various conditions, afflictions, and mental states. The most common, the Bach Flower Essences (because they were identified and categorized by Dr. Edward Bach), target specific mental and lifestyle states. Very expensive (about $10 per one-ounce bottle), they also contain a high percentage of alcohol.

For example, one Deva Flower Remedies product is labeled "Negative Influences" (to help the user overcome or withstand negativity), which includes walnut among its ingredients and is based in an 8 percent alcohol solution.

Interestingly, *Prevention Magazine's Health Book* also noted walnut as well as mustard as being good for fighting depression.

percent pure aloe. Don't take aloe if you're allergic to aspirin because it contains aspirin-like chemicals.

• **Anise seed,** an expectorant that also contains antiviral qualities, can be used to make an effective tea by steeping 1 to 2 teaspoons of seeds in 2 cups of boiling water for 15 minutes and then draining. It is said to help colds and flu.

• **Basil leaves,** fragrant as well as flavorful, rubbed on the skin acts as an insect repellent and is used as such in many parts of the world.

• **Bilberry,** also known as blueberry, is thought to help relieve menstrual problems by helping relax the uterine muscles and may also help some eye problems.

• **Black tea** in wet tea bags can be applied to sunburn. The tannic acids and other chemicals in black tea will help repair skin damage done by sunburn.

• **Calendula** ointment, used according to package directions, will help soothe inflammation in sunburn. Calendula is also very effective for minor skin burns and can be taken in the form of over-the-counter ointments because it has powerful antibacterial, antifungal, anti-inflammatory, and antiviral properties. Look for calendula as a main ingredient of the ointment. Use it according to the directions on the label.

• **Caraway** can be used in cooking beans and other flatulence-causing foods to minimize their gas-producing effect on the body.

• **Chamomile** tea is used for fungus infections. The sufferer is urged to drink a cup of chamomile tea made with a tea bag, then press the used tea bag on the fungus infection two to four times a day, or tape it on that area until the tea bag dries. This external use is for all areas of the body except the genitals or eyes or other especially sensitive areas.

Warning! Chamomile should not be used by anyone with hay fever–like allergies because it is related to the ragweed family.

• **Cinnamon** possesses stimulant, astringent, and carminative qualities, and supposedly helps lower sugar and hunger cravings as well helping the body fight viruses.

• **Citronella,** an Asian plant, can be used in a candle to ward off insects.

Warning! Avoid use of citronella directly on the skin. If taken internally, it can be immediately fatal and could probably have dangerous effects if absorbed through the skin.

• **Cloves** can be used in cooking beans and other flatulence-causing foods to minimize their gas-producing effect on the body.

• **Coltsfoot** (also called tussilago) is best taken in a tea made of 1 teaspoon per cup of boiling water steeped for fifteen minutes. It helps people suffering from coughs and other temporary respiratory problems. The tea can be drunk three times a day for coughs.

F.Y.I.

Cocoa butter is used for reducing and preventing skin wrinkles. It has the advantage of melting at body temperature so that it sends moisture into the skin, especially around the eyes. It's best obtained in moisturizers and other skin care products that have a large amount of cocoa butter listed in the ingredients.

Herbs to Avoid

Herbs that are particularly dangerous to everyone regardless of age and gender include the following:

• **Tea tree oil.** Many books and articles on herbs include information on oils taken from herbs plants, such as tea tree oil. This oil, however, has been found to be dangerous to the liver.

Some plant oils, such as lavender oil and oils from other aromatic plants, can be fatal if swallowed! The lesson here is, don't take any plant oil internally unless your physician assures you it is safe for ingestion in the body.

• **Pennyroyal.** This herb, used often for insect repelling, can damage the liver if absorbed in too great a quantity by the body's skin or if ingested. The same advice pertains to another but less common insect repelling herb, mountain mint.

• **Comfrey.** If taken internally, this dangerous herb can cause liver failure.

• **Sage.** Any more than a few leaves consumed in any form including teas, can cause convulsions for some people. Cooking with sage leaves for longer times is different—some poultry recipes call for up to fifteen leaves with no problems.

• **Dill** can be used in cooking cabbage and other flatulence-causing foods to minimize their gas-producing effect on the body.

• **Echinacea** is excellent for strengthening the immune system so the body can ward off viral and bacterial infections. Read the supplement label on a supplement bottle or ask a doctor for dosage advice on immune-strengthening purposes.

If you get one cold after another, you may try echinacea tablets in a two-weeks-on, one-week-off schedule. Ask your doctor for more details.

Echinacea tea or ½ teaspoon of echinacea in tincture form three times a day alternating with

• **Salicilin-containing herbs.** Willow bark, meadowsweet and cinnamon contain the same compounds as does aspirin. Don't take these or other herbs containing lesser amounts of salicilin (the main ingredient in aspirin), such as red pepper and cranberry, if you are allergic to aspirin. Consult a physician if you are not sure.

• **Herbal laxatives.** Senna and cascara sagrada are much too harsh for unsupervised use. Commercial diet teas, which have been known to be fatal because of their laxative properties, often contain these herbs.

• **Herbal energy jolters.** Ephedra (also known as ma huang), guaranine, ginseng, ginkgo, and the Chinese energy herbs have become famous for their rapid, amphetamine-like effects. But, like most substances that push a tired body and brain, they can elevate blood pressure to dangerous levels, overtax the heart, and possibly cause stroke or some other instantly debilitating or fatal effect. And these herbs can mask serious health conditions that are the cause of the low energy.

• **Sleep inducers.** Valerian and chamomile (especially valerian) can induce a sleep so deep in some people it interferes with their ability to breathe. A few drops of lavender oil in a hot bath is a safer way to induce relaxation and later lead to restful sleep.

lemon balm tea and *mint* tea drunk at least three times a day will all provide the body with antiviral compounds that fight a fever-blister outbreak.

Pressing the used tea bags against the fever blister until it is saturated or the tea bag dries provides even more benefits, or apply the tincture of echinacea directly to the fever blister. Echinacea in tincture form—60 drops three times a day for a week—is also said to help a sinus infection.

• **Elderberry** is said to work for flu because it contains antiviral agents that stop flu from moving into the lungs. Take it in a prepackaged tea form or make a tea with elderberry leaves, steeped for fifteen minutes in a cup of boiling water. It is also well known in Israel for curing colds, coughs, and fevers. It is sold over the counter in a product called Sambucol, which provides the proper dosage on the label.

• **Eucalyptus, licorice and slippery elm teas** will soothe inflamed sore throat tissues. Lozenges containing these ingredients may also help.

• **Fennel** is as useful as dill and cloves in minimizing the "gassy foods" effects on the body.

• **Fenugreek seeds** is used to help bowel problems. Taken 2 teaspoons a day with plenty of water (at least 8 glasses) should help heal constipation gently, without the harshness of laxatives. Fenugreek is also used for diarrhea when taken in capsules or tablets amounting to 1 or 2 teaspoons of the herb because it will add bulk to the stool and soothe the digestive tract.

• **Feverfew** has been shown to be so effective in about three-quarters of all users in warding off headaches and migraines that it has gained official recognition by the medical community in Europe. This herb can be taken in the form of capsules or teas, and never in more than the amount that would be available in three leaves of this herb.

People chew feverfew in England to prevent migraines.

Warning! Some people have a reaction to feverfew that takes the form of mouth sores. In this case, it must be discontinued. Also, it must never be taken by pregnant women. Taken over a long period of time it can produce an unwanted tranquilizing effect.

• **Flaxseed** is another excellent form of fiber that prevents, and helps heal, constipation without the harshness of laxatives. The oil in flaxseed is not toxic and is rich in valuable fatty solids.

Flaxseed can be taken in amounts of 1 to 3 tablespoons, two or three times a day, either alone (it has a nutty flavor) or added to foods. But it must be accompanied by at least ten glasses of water a day to provide the colon-cleaning benefits of fiber and not clog the digestive tract.

• **Ginger,** a fragrant and flavorful root, is used for colds and flu and provides many antiviral compounds. It works well when made into a tea. Ginger root, shredded and steeped in boiled water fifteen minutes, makes a useful ginger tea. Ginger tea, or ginger ale made with powdered ginger added to carbonated water and sweetened with honey, will fight a fever.

• **Garlic** is excellent for bronchitis because a chemical it releases in the body, allicin, is one of the most powerful broad-spectrum plant antiseptics. A woman who has suffered bouts of bronchitis all her life now treats it at the first sign of trouble. She consumes an *entire* head of garlic, peeled and roasted for one minute, followed by several glasses of water and a good night's sleep. Eating five or six garlic cloves a day helps build up antifungal power in the blood.

Garlic can also be used for insect stings in the following manner: Chop or blend onion or garlic until it is in a mashed form that can be applied to an insect sting and cover with a Band-Aid. Eating

The Wonder of Fruits

The fruit kingdom also yields healing properties. These common yet extraordinary fruits are just a few of nature's medicinal marvels.

• **Cranberry.** In juice form, dried (and sold in bulk like raisins or prunes), capsule form, or canned in jellied or relish forms, cranberry contains acids that act as antioxidants and prevent the spread of bacterial infection in the urinary tract. Because of this quality, cranberry helps prevent cystitis, a bacterial infection of the bladder. Cranberry also helps relieve the burning sensations and urge to urinate that accompany cystitis.

Cranberry, best taken in a concentrated form with a great deal of water and honey as a sweetener, contains aspirin-like ingredients that aid the discomfort of, and help lower, fever.

• **Grapefruit.** Eaten whole, not in juice form, in amounts of at least three a day, this citrus fruit provides enough modified citrus pectin (MCP) to help slow the growth of hard-to-treat cancer cells, such as those in melanoma, and specifically for men, in cases of prostate cancer.

The grapefruit must be consumed completely, except for the skin. The MCP is mostly contained in the white pulpy pith surrounding the fruit. If you don't want to eat that many grapefruits a day, MCP comes in powder or capsules allowing a daily supplement of 15 grams.

• **Grapes.** These can be mashed in a blender and applied to the face in a mask for 15 minutes then rinsed off to help prevent wrinkling. The alpha-hydroxy acids in the grapes will peel off the dead skin that can hasten wrinkling.

• **Papaya.** Not only a source of many antioxidants, papaya also contains papain, a protein-digesting enzyme so powerful it's used in many commercial meat tenderizers. Papaya in tablet form is a considerable digestive aid, especially for a mild upset stomach.

• **Pineapple.** This tropical fruit helps block the form of nitrosamines in the body that occur from eating cooked meats. Nitrosamines are carcinogenic. Pineapples are a good source of bromelain, an enzyme that helps excrete fats from the body. Dr. Andrew Weil claims that 200 to 400 milligrams of bromelain a day on a very empty stomach—at least two hours before a meal—will help heal sprains and black eyes.

raw or cooked garlic simultaneously will provide even more relief. Fresh parsley taken at the same time will mitigate the garlic smell.

It is important to drink several glasses of water after the garlic because garlic is drying to the nasal passages and also because the water will help decongest, and expel, germ-filled mucus.

• **Goldenseal,** an alterative, anti-inflammatory, diuretic, and laxative, among its many touted benefits, should be used prudently. You may try it in tincture form of 20 to 40 drops a day is used for a sinus infection.

It can provide such a strong antibiotic effect that it should be taken with acidophilus in some form to replenish the healthy bacteria that can be killed off by it. Because of this antibiotic effect, goldenseal should not be used for more than two weeks at a time.

• **Horehound** is well known for calming coughs. A tea purchased in prepackaged form or made of two teaspoons of this dried herb steeped fifteen minutes, drunk three times a day, should help.

• **Horse chestnut and witch hazel creams** will fight lines and wrinkles in the skin. These herbs were found in a major Japanese study to have sufficient antioxidant activity necessary for killing the free radicals that lead to wrinkling.

• **Lemon balm** tea drunk at least three times a day will provide the body with antiviral compounds that can fight a fever blister outbreak. It is also thought to be a mild sedative. Pressing the used tea bag against the fever blister until it is saturated or the tea bag dries provides even more benefits.

• **Licorice,** mentioned earlier as a mood elevator, also has properties to soothe inflamed sore throat tissues and relieve coughs. Lozenges containing this ingredient will also help.

• **Milk thistle** is said to detoxify the liver of all the polluting chemicals it processes for the body. You can buy this herb in prepackaged form (there are several good brands) and follow directions on the label.

• **Mint,** made into tea and drunk at least three times a day, will provide the body with antiviral compounds that fight fever blisters and flu. Pressing the used tea bag against the fever blister until it is saturated or the tea bag dries provides even more benefits. Peppermint and mint drunk three times a day, but not more often, will help heal the discomfort of indigestion.

• **Mullein** tea made the same way as the horehound tea and drunk three times a day is also recommended for calming coughs.

• **Nettle,** an alterative, antiseptic, and expectorant, is taken by many people to clear up hay fever symptoms. Take nettle in prepackaged form and follow the label—never overdose.

• **Pau d'arco,** from the inner bark of a Brazilian tree, is said to help reduce infections and fight cancer. An alternative health practitioner should be consulted before you take this herb.

• **Psyllium** in the form of seeds or powder or husks can be taken in amounts of 3 to 5 tablespoons a day mixed with juice or put in food—but must be accompanied by at least 12 ounces of water—can relieve constipation and unbind the intestines.

• **Red pepper** contains an aspirin-like ingredient that aids the discomfort of a fever and also helps lower it. It also is a help in overcoming obesity. A recent study in England found that adding ½ teaspoon of red pepper to food served at a meal can increase the body's metabolic rate as much as 25 percent without causing the same kind of dangerous side effects as ephedra.

Warning! The ½ teaspoon of red-pepper-at-meals dosage should never be exceeded and should be accompanied by drinking large amounts of water.

• **Saw palmetto** is said to be excellent for prostate problems.

• **Slippery elm** helps coughs by coating the throat and mouth with a soothing, mucilage-like substance found in the slippery elm tree bark. It is best taken in throat lozenges that will provide throat coating.

Slippery elm has been declared a safe and effective cough suppressant by the Food and Drug Administration. A tea made with slippery elm will soothe inflamed throat tissues. Lozenges containing this ingredient will also help.

• **Tea tree oil** contains a strong antiseptic called terpinen. It should be kept on hand for external application to cuts and other abrasions of the skin.

• **Witch hazel** soothes skin irritation when applied externally and will fight wrinkles.

THE BOTTOM LINE

The public has become increasingly concerned about the use of ever-stronger antibiotics, prescription pharmaceuticals, and other medications, and of the unwanted side effects from these. Many people now consider herbs and plant therapies as alternative routes to maintaining and restoring health. The wide variety of plant sources available and their array of healing properties make them worth investigating.

There are many safe plants and herbs to take—garlic, echinacea, cranberry, and more. Many others should be taken only with advice and supervision from your health-care practitioner.

Healing Supplements for Women

Women have special health problems that can be relieved by healing supplements in a number of ways. Premenstrual syndrome, menopausal symptoms, vulnerability to breast cancer, and what our grandmothers and great grandmothers referred to as "delicate nerves" (female depression) are problems most women must deal with not just once but several times during their lives.

Until recently, the amount of suffering women were expected to put up with in silence because of these physical problems was impossible to measure. Now, finally, women's health issues are being taken seriously—not just by the medical establishment but by women themselves. They're beginning to find that alternative methods to traditional medicines—such as the healing supplements covered in this chapter—can ease their health problems in ways that are good for them.

Women and Hormones

Women are physically different from men in an especially, if almost obviously, important way: Women have their own complicated hormone systems that make childbearing possible. These female hormones work all during the childbearing years, not just in pregnancy and lactation, lessening to menopause.

Of the possibly one hundred hormones in the human body—not all of them have been pinpointed yet, according to the standard medical references—women are most affected by two: estrogen and progesterone, the ovarian hormones.

Their overall functions, simply:

• **Estrogen** influences the formation of secondary sex characteristics, such as breasts, a higher-register female voice, limits body hair and is a key hormone involved in ovulation.

• **Progesterone** allows the uterus to prepare for pregnancy—and to carry the baby to term.

When a woman menstruates, it usually indicates that there's no pregnancy as the unfertilized egg is sloughed off every month. This process may be nature's cycle, but with it, hormone changes may unsettle a woman's sense of well-being in a number of ways. For one, a few million women may sing the blues every menstrual cycle because of a condition called PMS, or premenstrual syndrome.

Inside Premenstrual Syndrome

Renee, a bright, young woman of thirty years with a job of associate fashion editor at a national magazine, summed up her problems with PMS this way: "I lose all my confidence when I'm about to get my period."

Many medical experts, most of them male, once insisted that premenstrual symptoms, especially mood swings, were either imagined or the result of life crises, such as a romance going badly.

F.Y.I.

Some women in their forties can confuse the symptoms of PMS with the onset of menopause. The symptoms of menopause are menstrual irregularity, depression, bloating, and even night sweats. This usually occurs in women who have gained a lot of weight or who have had a tubal ligation (sterilization)—and chances are that what they're experiencing is actually PMS, not the first signs of menopause.

Research on this condition in the last ten years has shown that PMS is real—and its unpleasant symptoms can be alleviated or reversed.

Almost all nonmenopausal women are affected by PMS, to some degree. Symptoms cover the range from loss of intellectual acuity, to nausea, bloating, uncontrollable appetite (especially for salt, sugar, and chocolate), to painful cramps.

Women through the centuries have attempted to self-treat, following the advice of other women as to "what works" to end the cramps, for example—from sitting in very hot bath water spiked with an infusion of a seaweed solution to drinking icy ginger ale.

No one knows what the real cause—or cure—of PMS is. But most doctors suspect it is a hormonal imbalance that comes on with menstruation. One traditional or allopathic medical treatment includes birth control pills, which your doctor must prescribe, if she thinks it's right for you. Often the estrogen in these pills can regulate a hormone imbalance causing PMS. But they don't work for every woman.

Many alternative physicians have long been recommending healing supplements to help alleviate, if not cure, PMS.

Aids for Relieving PMS

There are a number of routes PMS sufferers can take to free themselves of unpleasant premenstrual ups and downs. Unless otherwise indicated, these supplements should be taken every day in addition to the regular supplement schedule, not just when PMS symptoms occur. All of them can be found in specialty vitamin shops or health food stores.

• **Vitamins** . Extra B$_6$, taken in a dosage of 100 milligrams twice a day when symptoms first occur until they subside and vitamin E in the standard dosage of 200 IUs.

A study of the effect of vitamin B$_6$ showed that a dosage of 500 milligrams a day provided twenty-one out of twenty-five women with noticeable relief from PMS symptoms, and did not cause as much fluid retention, weight gain, and acne outbreaks. However, experts agree that doses over 200 milligrams a day of B$_6$ can cause problems with the nervous system. Staying with 100 milligrams of B$_6$ a day is safest and probably just as effective. Some gynecologists may also suggest cutting down salt intake, too.

• **Dong quai.** In his best-selling book *Spontaneous Healing*, Dr. Andrew Weil recommends a dropperful of tincture of dong quai in water twice a day for women who suffer extreme PMS and other female problems, such as irregular menstruation. This herb could be taken for six to eight weeks before results can be expected. It should be avoided by any woman with elevated blood pressure.

• **Gamma linoleic acid.** GLA taken in dosages of 500 milligrams twice a day also helps PMS. Best sources are in capsule form found in health food stores. GLA is also found in evening primrose oil and black currant oil capsules.

• **Vitus agnus-castus.** Also called chasteberry, this is an herb that may help regulate the female reproductive cycle. It can be taken in capsule form or tincture—that is, one dropperful of the tincture in water or two capsules twice a day.

One side effect of this herb can be repression of the sexual drive. It's called chasteberry (or monk's herb) for a reason—it has been used in convents and monasteries throughout history.

• **Black haw bark.** This may reduce the pain of menstrual cramps by relaxing the uterus and calming muscle spasms. Experts recommend 20 drops of the tincture, up to three times a day.

• **Raspberry leaf.** Raspberry leaf tea, also said to relax the uterus thereby reducing cramps, is made by using 1 teaspoon of raspberry leaf to 1 cup of boiling water. Let it steep a few minutes and cool down a bit before drinking. It can be drunk when needed, up to four cups a day.

• **Pycnogenol.** These super antioxidants described in chapter 4 were shown to lessen menstrual cramping significantly after being taken for four months.

• **Kava-kava root.** Dr. Alex Cadoux call this "one of the most successful instant mood lifters" he knows of to aid women suffering from PMS tension, irritability, and mood swings. This "herb" is best taken in capsule form.

Dealing with Menopause

With the cessation of ovulation, menopause begins. At first, the only symptom may be irregular periods. This is followed by sleeplessness, hot flashes, vaginal dryness, even osteoporosis (down the line, if she is so predisposed), and other symptoms.

Because menopause is caused by declining estrogen production, its symptoms are often alleviated with hormone replacement therapy (HRT). Medical doctors debate the "yes, take hormones; no, don't take hormones" ideologies, but generally, those who are *pro* hormone agree that whether or not a woman's had a hysterectomy, HRT is probably a good idea.

HRT involves taking estrogen and progesterone—how many years depends on the woman, her response to the hormones, her doctor's advice, research on hormone replacement, and its effects on the body and any new decision she may make about her health care. Progesterone is thought to be a necessary complement to estrogen because it mitigates the cancer-causing effects of estrogen.

The difference between hormones produced naturally by women *before* menopause and those taken in supplement form *after* menopause is this: Naturally produced estrogen is channeled directly from the ovaries into the bloodstream to circulate throughout the body. Hormones taken orally must go through the stomach and liver and then be broken down into other compounds—a step that may affect the body in ways not yet known.

HRT, some research says, helps prevent heart disease, osteoporosis, some types of mental deterioration, and more, but perhaps increase the risk of some cancers. HRT is also said to delay the onset of Alzheimer's disease, if the woman in question has the predisposition to it.

According to an article in the Harvard Medical School publication *Women's Health Watch*, it is

F.Y.I.

A vegetarian diet can help fight hot flashes and other symptoms of menopause. The reason? Many plant foods, most notably soy bean, contain natural components called phytoestrogens. These plant chemicals, especially isoflavones, provide many of the same benefits to the body as does natural estrogen.

SMART MONEY

Expert Alex Cadoux, M.D., says that women who can't tolerate estrogen replacement therapy should ask their doctors about taking natural estrogen, in the forms of *estral* and *estradiol*. It's not that synthetic hormones, which are used in HRT, are necessarily bad, Dr. Cadoux says, "but they might contain substances that are not well tolerated by some women."

the loss of the estradiol form of estrogen that causes menopausal symptoms, rather than two other estrogens women also produce—estriol and estrone. In fact, there is now available a plant-based estrogen being sold under the trade name Estrace, which provides estradiol.

However, current research shows that oral HRT is largely effective in curing most menopausal symptoms. Hormones provided by a transdermal patch applied to the skin would bypass the liver. Your doctor can provide more information on these choices.

Plant-based progesterone, Prometrium is just one brand of it, has been found to be more beneficial to women than the synthetic forms of it. Plant-based progesterone is especially good for raising HDL (good cholesterol levels) and for avoiding side effects caused by synthetic progesterone. Again, seek your doctor's advice about plant-based progesterone to know it they're right for you.

Many menopausal women begin on mainstream HRT and switch to nonpharmaceutical alternatives. One reason is to avoid osteoporosis. And experts say that HRT tends to increase a woman's daily requirement of vitamin B_6 to about 250 milligrams a day.

Supplements to Relieve Menopausal Symptoms

For women who do not want to take any hormones after menopause, there are supplements available. The following are said to help:

• **Black cohosh.** Introduced to Europeans by Native Americans in the northeastern parts of the continent, black cohosh is the primary ingredient in a product called Lydia Pinkham's Vegetable Compound, long used in this country for "female complaints" and still produced in Europe. Black cohosh is not to be confused with *blue* cohosh, an herb that can stimulate the uterus to contract and bring on menstruation and miscarriage.

• **Chasteberry.** Also mentioned as an aid for relieving PMS symptoms, chasteberry can be taken in capsule form.

Any multi-ingredient menopause-aid capsule or tablet may contain other herbs, but should list at least one of these two herbs as a main ingredient.

• **Melatonin.** One of the worst symptoms of menopause, which gets the least amount of attention, is insomnia. One of the reasons for sleeplessness relate directly to the cessation of estrogen production and other hormones. But it could also be the coincidental loss of melatonin production at the age of menopause.

Melatonin is a hormone secreted by the pituitary gland that, among its other functions, causes sleep and wakefulness. If taken in supplement form, melatonin can make the body think it's time to feel drowsy and sleep—hence, its common use in combating jet lag. The pituitary gland stops producing this hormone as we age and lose our ability to reproduce.

In his book about melatonin, Dr. Stephen J. Bock, M.D., says this hormone is a powerful antioxidant that fights free radicals. Chronic ill-

F.Y.I.

Hot *flashes* and hot *flushes* are not the same thing, although both are directly related to the dilation and constriction of blood vessels. A hot *flash* results in the sudden feeling of heat throughout the body, with no sweating or skin sensation. A hot *flush* occurs from increased blood flow to the skin and vital organs and begins with profuse sweating, and a reddening of the skin on the face, neck, and upper chest.

Is Melatonin for You?

When melatonin first became touted as a miracle drug a few years ago, the suggested dosage was about 3 milligrams per night. Then it was found that 3 milligrams made many users sleep so heavily that they were groggy and felt hung over the next day. Therefore, 1 milligram may be all that's necessary for to sleep soundly and feel alert the next day.

Important! Melatonin is a hormone, not a nutrient like vitamin C whose excess in the body is easily excreted. Be smart! Ask your doctor about whether it's right for you, especially if you're taking any medication.

ness, sleep disturbances, and "phase shift symptoms"—such as feeling sleepy during the day and alert and awake at night—are signs of melatonin deprivation. Dr. Bock believes that supplements of melatonin can restore some of its benefits to the body. Some experts believe that synthetic melatonin, which is sold without a prescription at vitamin shops, drug stores, and health food stores, is safe when taken judiciously. A dosage of 1 to 2 grams taken one hour before bedtime is considered typical. If sleepiness is immediate, take your melatonin closer to actual bedtime.

Menopause and Disorientation

A sense of losing control—that is, as if you're "not feeling like yourself" or "feeling hypersensitive"—affects women dealing with PMS differently than it does for those coping with the changes in menopause.

The sense of disorientation in PMS lasts just a few days—not that it may not be intense. With menopause, it often does not go away in a few days, and, too, it may become more serious if not kept in check. Sudden mood swings, an inability to make up one's mind, and other unpredictable behaviors may occur.

Here's an example: Carrie, a twenty-five-year-old nurse practitioner, looked forward to visiting her parents after not seeing them for a year. She and her mother had always been extremely close, and her mother had been very supportive of Carrie's career. But, Carrie found that on this visit, she had a different mother.

"This woman, who had raised six children successfully and operated a huge house all her married life, now couldn't set up a lunch date. She'd decide on one coffee shop, then for some reason change the time, the place and then the day, every two minutes. She'd alienated every friend she had and spent her days either cleaning obsessively or looking out the window motionless for hours. Yet she refused estrogen therapy or anything else that might help!"

In instances such as these a physician's care is crucial. The woman who will not see a doctor for any reason might want to try St. John's Wort three times a day, or melatonin at night, and prepackaged herb blends made for menopausal symptoms (containing a sufficient amount of black cohosh and chasteberry) every day.

Menopause and Memory Problems

Many experts believe that memory loss associated with menopause may be due to a vitamin-B deficiency, particularly deficiency of B_{12}, but also of folic acid and thiamin. The B-complex vitamins are thus particularly important for women at this time of their lives.

Other more serious memory loss and mental deterioration may be diagnosed as Alzheimer's disease, a hereditary disorder, which also affects men. Women who fear Alzheimer's disease after menopause should consider the following:

• Estrogen replacement therapy has been found to help forestall the onset of Alzheimer's in some women who are at risk of this disease.

• People with Alzheimer's disease were found to have elevated levels of aluminum in their brains. Fluoride competes with aluminum for receptor sites in the brain, so the more receptor sites that are found by fluoride, the better. Thus, fluoride-treated water and toothpastes may possibly be valuable for helping lessen the symptoms of Alzheimer's over a long range as well as prevent cavities in the teeth.

Healing Supplements for Breast Health

American women have a 500 percent greater incidence of breast cancer than women of any other culture. It has been traced, in part, to a high-fat diet. Eating fats, especially with protein that has been charred in the cooking, and *being* overweight are factors that raise the risks for getting breast cancer.

Doctors now think the results of a test for selenium and vitamin E levels in a woman's body can be an important indicator for breast disease. If they find *low levels* of both nutrients, it may indicate a higher risk for breast cancer and/or fibrocystic breast disease, in which, in the latter, nonmalignant cysts form in breast tissue.

The following healing supplements have been looked to for their positive effects.

• **Vitamin E and Selenium.** These were found to relieve breast tenderness during bouts of PMS as well as reduction of and relief from certain breast cysts. Some doctors believe 600 micrograms of selenium a day is a safe and effective possible means of boosting immune system function and possibly reducing cancer risks. Check with your physician to be sure.

Warning! Never exceed 200 micrograms of selenium a day; more than this amount is toxic.

• **Beta-carotene.** In studies with European women, women who tested *less* than 3,300 IUs of beta-carotene, a known antioxidant, in the body were found to have an *increased* risk of breast cancer. Other researchers are not convinced that beta-carotene is a real cancer fighter. Other experts differ on *how* beta-carotene is best obtained and utilized most efficiently—from supplements or food.

Until more evidence is in, it's best to get beta-carotene from fruits and vegetables rich in it, for example, carrots, cantaloupes, and yellow or orange squashes. One cup of broccoli, five carrots, and three mugs of green tea provides sufficient protective amounts of beta-carotene a day for breast health. Microwaving all or some of the vegetables will release some antioxidants that are not gotten from eating raw vegetables.

• **Milk thistle.** Women worried about the effects of radiation from mammograms may want to try milk thistle, an herb whose antioxidant properties have proven effective against X-ray radiation.

F.Y.I.

An epidemic in the United States, breast cancer strikes one of every eight American women. Some estimates claim the prevalence closer to one in seven women.

Many believe this herb is best taken in a standardized extract of milk thistle seed called silymarin, available at health food stores.

• **Omega-3 fatty acids.** Not all fats are bad for breast health. The omega-3 fatty acids, especially gamma linolenic acid found in evening primrose oil, are available in capsule form and have been found useful in reducing some nonmalignant breast tumors.

• **Ginseng.** Scientists abroad have found that pure ginseng reduced breast tumors in mice. This finding suggests that, if no hypertension is present, ginseng supplements might help lessen breast tumor growth in women.

What Women Should Know about Osteoporosis

This condition, often hereditary, usually appears in older women because of menopause-induced bone loss. Women on hormone replacement therapy need only about 1,000 milligrams of calcium a day because the estrogen has been found to forestall osteoporosis. Otherwise, a high intake of calcium of about 1,500 milligrams a day can often help prevent osteoporosis.

Never exceed 1,500 milligrams a day of calcium, if you do not take estrogen, or 1,000 milligrams a day with hormone replacement. Too much calcium can cause magnesium deficiencies—calcium and magnesium compete for the same receptor sites in the body—and lead to deposits that might cause arthritis or, in those prone to the condition, kidney stones.

Acidophilus: The Cure for Yeast Infections

Acidophilus bacteria are responsible for souring milk and providing "good bacteria" in the intestines that make sure nutrients in food are being absorbed. These beneficial bacteria fight infection and prevent vaginal dryness and yeast infections. An insufficient amount of stomach acids can lead to all sorts of problems, including female hair loss. This condition means nutrients are not being broken down by and absorbed into the body. Regular daily doses of acidophilus in yogurt or supplement form can do a great deal to maintain health.

Antibiotics often kill off these good bacteria, which is why women who take antibiotics are so prone to yeast infections.

In the event a yeast infection occurs: You may take acidophilus in capsule, powder, or liquid form—all of which are available at health food stores and vitamin shops. It's best to get the refrigerated variety. Refrigeration ensures that the live bacteria cultures in the acidophilus are still alive.

Sufferers of yeast infections are also urged to buy good-quality yogurt, preferably a brand from the health food store rather than a supermarket brand. Some of this yogurt can even be applied to a sanitary napkin and then worn so the yogurt is

F.Y.I.

There's an added benefit to taking calcium discovered in a study with men: Ten healthy men at high risk for colon cancer were given calcium carbonate supplements of 1,250 milligrams daily. These supplements were believed to have reversed the abnormal colon cell growths found in these men to a state less likely to turn into cancer.

The extra calcium is thought to bind with extra fats in the colon and carry them out with feces. These fats can become a source of dangerous oxidants when they putrefy in the intestines—a likely occurrence if fats are not expelled before putrefaction occurs.

in contact with the vaginal area. Fresh applications on fresh pads could be tried every few hours if possible.

Some alternate-care doctors may even advise insertion of an acidophilus capsule into the vagina, as easily as if it were a tampon. Talk to your gynecologist about this first.

Liquid acidophilus is very powerful and pleasant tasting and can be drunk a few tablespoons at a time every few hours, or after meals. In his book on alternate health care, Dr. Andrew Weil recommends that women with a yeast infection eat a clove of raw garlic with a few sprigs of parsley every day until the infection is cleared up. Garlic contains powerful antifungal properties that are especially useful for alleviating yeast infections.

Note: Acidophilus may be labeled with a product brand name. Read the ingredient label: As long as the product has live cultures and its expiration date has not been reached—*be sure to check the container!*—and is kept refrigerated, it will be fine.

Homeopathy and Women's Health

Homeopathy is a kind of medical practice that has researched and created unique, individualized medicines that seem uniquely suited to the complex female body.

Homeopathy is based on the law of similars, or "like curing like." This philosophy of curing has been in existence from the time of Hippocrates in the Golden Age of Greece and has been practiced in many cultures since. The very effective notion of immunization by vaccination is based on this idea, but before this, a German physician, Samuel Hahnemann (1755–1843), found through careful experimentation that the law of similars could be used to treat many diseases and conditions. Successes followed.

Homeopathy was found to be much more successful at curing victims of cholera and other diseases that were sweeping Europe and America in the late nineteenth and early twentieth centuries than did traditional (that is, allopathic) medicine. But a kind of medical war ensued wherein more established medical doctors managed to drive homeopathy to near extinction in the mid-twentieth century. But homeopathy has survived. With the rise of interest in alternative health care, the specialty has since proved so valuable it has begun to flourish again.

How to Find a Homeopathic Specialist

The best way to use homeopathic remedies is with the guidance of a physician who has studied, and been certified in, the use of homeopathy.

Many branches of alternative medicine license their own members as homeopathic physicians. Medical doctors interested in becoming a homeopathic physician can study with another homeopathic physician or in a medical college abroad that offers courses in this specialty. Thanks to the repression of homeopathy in this country earlier in the century, there are still no medical colleges in this country that offer such courses.

F.Y.I.

A German physician, Dr. Samuel Hahnemann, developed the practice of homeopathy in the early nineteenth century as a response to the orthodox methods of treatment of the day—bloodletting being one of the more popular practices.

The best bet, though, is to find a homeopathic physician who will prescribe for, and monitor, your condition. You can find an M.D. licensed to practice homeopathy listed in the Yellow Pages, or trying searching the Internet using *Homeopathy* as the search word.

Buying Homeopathic Remedies

You can find several homeopathic remedies right on the shelves of some health food stores or vitamin shops that claim to treat conditions that tend to be especially common in women— fatigue, nervousness, tension, menopausal symptoms, and PMS, to name just a few.

Hyland's Nerve Tonic, for instance, says it addresses "nervous exhaustion" and "insomnia." This tonic contains calcium, potassium, magnesium, and iron, all listed with a "3X" after the ingredient.

To understand what the 3X code means, let's examine another example of a homeopathic remedy sold by Hyland's that is named Cough and Sore Throat Phosphorus 30X. According to homeopathic physician Dr. Henry Glover, M.D., the 30X means this medicine contains virtually no material phosphorus at all, just "the energy of phosphorus." This statement is not as transcendental as it sounds. Here's what Dr. Glover means:

The original ingredient, phosphorus, was diluted in the amount of 1 drop to 99 drops of water and pounded (or secussed) 100 times. One drop of that dilution was then subjected to the same treatment. One drop of that second dilution was then given the same treatment. A drop of that third fluid was diluted and secussed, and the same process repeated until thirty successive dilutions had been made—hence the 30X.

The Cough and Sore Throat Phosphorus 30X potion could be said to contain a relatively small essence of the original substance in it. On the other hand, the 3X in the Nerve Tonic means its ingredients have been subjected to just three dilutions and are present in much larger amounts.

Some homeopathic medicines go up to 50,000 C, which Dr. Glover compares to a "drop [of the original material substance] in a lake." In this case, there is nothing but the essence, or energy, from the original substance left. Why go to all that trouble? Because in homeopathy, less is often more, and the higher numbers can mean much more potent effects.

The idea is not that a specific ingredient in a homeopathic remedy may cure a specific symptom—thus, one should not deduce that phosphorus is good for coughs and sore throats—but that the ingredient in a particular dilution has shown to be effective in helping the body achieve a *balance* that allows it to deal with that condition.

Unlike allopathic medicine that often zeros in on a symptom, leaves the rest of the body vulnerable to side effects, and suppresses the real cause of the condition deep into the tissues where it continues to exist, homeopathic medicine considers the whole body to be out of balance and in need of medicine that will most effectively redress the imbalance.

THE BOTTOM LINE

This chapter looked at a few of the health issues women may deal with, at least for some time, on a continuous basis—such as premenstrual syndrome, yeast infections, menopause, and even osteoporosis. It compared traditional hormone therapy for both PMS and menopause with alternate-treatment suggestions—herbal blends, vitamin and minerals, and homeopathic remedies—to help relieve the symptoms of these conditions, such as mood swings, irritability, water retention, fatigue, hot flushes, and more. Hopefully this chapter provided some alternatives and reassuring advice.

What to Expect from Homeopathic Remedies

The homeopathic physician chooses medicines he or she uses from the bible of homeopaths, the *Materia Medica,* and often allows several weeks to see if the medicines work. Very often, Dr. Glover says, the patient won't experience a miraculous, instant change. She will eventually realize that she feels a lot better—often in ways she didn't expect. That means the bodily imbalance has been corrected and her overall health improved.

Which is not to say the results of homeopathic medicines cannot be instant and dramatic, some are. Hyland's Cough and Sore Throat can act as an instant expectorant and many homeopathic asthma remedies are equally rapid in effect. It's just that homeopathic physicians are determined not to make instant effectiveness claims.

Yet this kind of medicine does work or it wouldn't be used so extensively in Europe and would not be growing in use in this country. Thus, women who try these prepackaged homeopathic remedies should give them a while to work and not expect instant relief—although they may get them.

Warning! Dr. Glover says that homeopathic medicines can, like all medicines, be harmful in larger-than-recommended doses. Take them exactly as instructed on the formula.

A Final Note

The purpose of this book has been to introduce readers who've been skeptical or confused about supplements to their possible health benefits, with the hope that the information herein has provided food for thought. The approach of this Smart Guide has been to impart information and describe the research being done on supplements, and to be conservative in dosage recommendations.

Readers who want to begin a supplement program are advised to proceed cautiously and to speak to a licensed, trustworthy, and knowledgeable health-care professional. Those readers who want to know more about nutrient and nonnutrient supplements and what they can and cannot do should do some reading on their own—whether it is a professional health-care journal; a magazine whose reporters stay on the cutting edge of the benefits of nutrients, such as *Muscle & Fitness, Prevention,* or *Men's Health;* or connecting through the Internet with organizations that specialize in alternate (or complementary) health care.

Most of all, the reader is cautioned not to believe all authorities implicitly, no matter how famous. Rather, readers should do comparative research to learn what others have to say.

Rest assured that the field of healing supplements is a fascinating one—it's all about you and the body you were born with and how to live in it longer. Great, good health to you!

You may email your comments about this book to: ricker@u.arizona.edu.

—Ruth A. Ricker, Ph.D.

Appendix

This appendix provides a chapter-by-chapter cross-referencing of all of the supplements noted in the book and their links to particular physical ailments and conditions. Unless noted otherwise, a given supplement is generally considered to have a positive effect on the condition listed.

Thus, more than one chapter will have an entry for, say, cancer, because this guide looks at a full range of possible healing supplements for this disease.

Note: "All antioxidants" is a frequent supplement listing in chapter 4 because space precludes the listing of all of the antioxidant supplements.

Warning! None of these supplements should be considered as substitutes for medical advice or treatment by your health-care practitioner. These supplements are not to be regarded as cures for the conditions under which they are listed nor is any supplement to be regarded as exclusively suitable for the condition under which it is listed in this book.

Chapter 1: All about Healing Supplements

Asthma	Vitamins, antioxidants, echinacea
Baldness and graying of hair	Biotin
Birth defects	Folic acid; biotin
Immune system health	Echinacea
Depression	Thiamin

Cataracts	Thiamin
Cell damage caused by smoking	Cobalamin
Cholesterol level regulation	Niacin
Fragile fingernails	Biotin
Health of the heart	Pyridoxine
Health of the immune system in the elderly	Pyridoxine
Infection in the body	Pyridoxine
Mental confusion (especially in the elderly)	Cobalamin
Mental and physical stress on the entire body	Riboflavin
Nicotine and alcohol damage to esophagus	Riboflavin
Rectal and colon cancer	Folic acid; biotin

Chapter 2: The Healing Power of Vitamins

Absorption of other B vitamins	Biotin
Acne	Vitamin A
Aging signs	Vitamins E, A
Anemia, pernicious	Cobalamin, pyridoxine
Appetite loss	Folic acid
Arterial plaque	Vitamins C, A
Birth defects	Folic acid
Bleeding gums	Vitamin C, bioflavonoids
Blood clot thinning	Vitamins C, E
Bone deformation	Vitamin D
Bone health	Vitamin D

Bowel disease	Vitamin D
Breast cancer prevention	Vitamin D
Breast cysts	Vitamin E
Calcium absorption	Vitamin D
Cancer	Vitamins C, A, E
Carcinogenic reduction	Vitamins C, E nitrosamines
Cataract prevention	Thiamin
Cellular damage	Vitamin E
Cervical cancer	Cobalamin
Cervical dysplasia	Cobalamin
Cholesterol regulation	Pantothenic acid
Colon cancer	Vitamin D
Cognitive ability	Vitamin C, niacin
Depression	Folic acid
Eczema	Cobalamin
Energy increase	Cobalamin, niacin
Esophagus cancer prevention	Vitamin A, riboflavin
Eye membrane health	Vitamin A
Fingernails	Biotin
Hair loss	Biotin
Hair graying	Pantothenic acid
High blood pressure	Vitamins D, C
HDL level in the blood	Vitamin E
Heart disease	Vitamin C
Immune system strengthening	Vitamin C, pyridoxine
Mental confusion	Cobalamin, thiamine, niacin
Memory loss	Vitamins E, B-complex, niacin
Metabolism booster	Pyridoxine

Mood and sleep disorder	Folic acid
Oxygenation of the blood	Vitamin E
Platelet clumping	Vitamin E
Pollution damage	Vitamin A, riboflavin
Red blood cell deficiency	Riboflavin
Rheumatoid arthritis	Vitamin D
Skin dryness	Vitamins A, E
Sleep disorder	Vitamin E, pyridoxine
Stroke and thromboembolism prevention	Vitamins A, D, E, B-complex; niacin
Stress effects	Vitamins E, B-complex, C
Tooth decay	Vitamin D
Tuberculosis	Vitamin D
Varicose veins	Vitamin C, bioflavonoids
Viruses	Vitamin C
Vision	Riboflavin
Wounds	Vitamin C

Chapter 3: The Healing Power of Minerals

Alzheimer's disease, possible cause of	Aluminum, silicon
Amino acid absorption	Sulfur
Anemia	Iron
Antioxidant protection in cells	Selenium
Anxiety	Calcium
Arthritis	Calcium, sulfur

Attention span	Phosphorus
Birth defects, possible cause of	Lead
Blood pressure regulation	Potassium
Bone health	Calcium, phosphorus, magnesium, boron, manganese, silicon
Bone deformation	Manganese
Blood sugar regulation	Chromium
Brain function	Chloride
Cardiovascular disease	Calcium
Cancer, possible healing of	Selenium, vanadium
Cancer of the possible cure of	Molybdenum
Cardiovascular disease	Magnesium
Cancer, possible cause of	Chromium; too much iron; too much zinc; fluoride
Chronic fatigue syndrome	Magnesium
Cartilage health	Manganese
Cholesterol regulation	Chloride
Cholesterol level rise, possible cause of	Zinc
Chronic fatigue	Selenium
Colon cancer	Calcium
Complexion drabness, cause of	Sulfur
Constipation, possible cause of	Iron
Depression	Calcium, lithium
Digestive system health	Iron
Dwarfism, possible prevention of	Zinc

E vitamin absorption, possible barrier to	Iron
E vitamin helper	Selenium
Fatigue	Phosphorus
Fluid balance	Potassium, sodium
Free radical fighting	Copper
Gastroenteritis, possible cause of	Fluoride
Goiter	Iodine
Growth of the body	Zinc
Hair health	Silicon
Heart attack, possible cause of	Too much iron
Heart muscle disease	Selenium
Heat exhaustion sodium	Potassium,
High blood pressure	Calcium
High cholesterol	Calcium level
Hydrochloric	Chloride acid production
Hyperactivity	Calcium
Immune system health	Zinc
Infection healing	Zinc
Infertility	Selenium
Inflammatory disease	Manganese
Insulin regulation	Chromium
Iron absorption	Copper, vitamin C
Kidney stones	Magnesium
Leg cramps	Calcium
Liver function	Chloride
Low energy	Manganese
Metabolism	Phosphorus, magnesium, zinc

Muscle cramps	Calcium
Muscle health	Calcium, magnesium
Nerve health	Calcium, copper
Obesity	Vanadium
Osteoporosis	Fluorine, manganese
Oxygen transport and storage	Iron, molybdenum
Periodontal disease	Calcium
Potassium balance	Calcium
Restless leg	Calcium complaint
Tooth decay	Fluoride
Prostate cancer, possible cause of	Cadmium
Shock	Potassium, chloride, sodium
Skin health	Silicon
Sterility	Manganese
Strains and sprains	Manganese
Teeth health	Calcium, fluoride
Thyroid health	Iodine
Ulcers	Zinc
Undeveloped genitals	Zinc
Wound healing	Zinc

Chapter 4: The Healing Antioxidants

Asthma	Vitamin C (with flavonoids, hesperidin, rutin)
Allergies	Vitamins A, C (with flavonoids, hesperidin, rutin), E; selenium
Atherosclerosis	Vitamins A, C (with flavonoids, hesperidin, rutin), E; selenium

Bacterial and viral	Vitamins A, C infections (with flavonoids, hesperidin, rutin); selenium, garlic
Bronchitis	Garlic; vitamins A, C (with flavonoids, hesperidin, rutin); selenium; echinacea
Cancer	Vitamins A, C (with flavonoids, hesperidin, rutin); beta–carotene–rich foods; fiber; glutathione or gluta-mine; selenium; tannins in green tea; milk thistle
Colds	Echinacea, vitamin C (with flavonoids, hesperidin, and rutin), selenium, zinc
Eye and vision	Vitamin C (with problems flavonoids, rutin, hesperidin), selenium
Problems of aging	All antioxidants, beta-carotene rich foods, coenzyme Q10, glutathione or glutamine, selenium
Damaged DNA	Vitamin E
Depression	All antioxidants
Free radical destruction	All antioxidants, astralagus, ginseng
Fungal infection	Garlic, selenium sulfide
Heart disease	Antioxidants, beta-carotene–rich foods, coenzyme Q10, red wine extract
High serum cholesterol	Red wine extract
Histamine	Vitamin C (with overproduction flavonoids, rutin, hesperidin)
Infected tissues and wounds	Vitamins A, C (preferably with flavonoids, rutin, hesperidin) Mental problems
hesperidin) Mental problems	Coenzyme Q10, ginkgo biloba, selenium, vitamin E

Radiation damage	Vitamin C (with prevention flavonoids, rutin, hesperidin), tannin (in green tea)
Smoker's chalky white skin	All antioxidants
Sore throat	Vitamin C (with flavonoids, hesperidin, rutin), echinacea, zinc
Vascular disease	Coenzyme Q10, vitamin E, selenium

Chapter 5: Amino Acids

Allergies	Tyrosine
Aging signs and diseases	Glutamine
Acetaminophen overdose	Cysteine
Cancer	Cysteine, glutamine, taurine
Creation of creatine	Arginine, methionine, glycine
Food cravings	Phenylalanine
Growth of the body	Histidine, isoleucine (especially for infant growth), leucine, lysine, threonine, valine, tyrosine
Heart disease predictor	Homocysteine
Heart disease helper	Carnitine, methionine
Heavy metals detoxification	Methionine
Infections	Cysteine
Insomnia, lack of happiness	Tryptophan, tyrosine, methionine, carnitine, phenylalanine
Intellectual ability	Histidine, glycine, glutamic acid
Kidney and liver function	Alanine, arginine
Metabolism	Methionine with pyridoxine, phenylalanine, carnitine, aspartic acid, valine, taurine

Neurotransmitter firing in the brain	Choline
Nutrient absorption	Proline, serine
Pica (eating clay and other inorganic minerals)	Mineral and vitamin complex
Radiation effects	Glutamine
Selenium absorption	Methionine and pyridoxine

Chapter 6: The Healing Plant Kingdom

Allergies	Nettles
Antidepressant	St. John's Wort
Aspirin effects	Cinnamon, cranberry, meadowsweet, red pepper, willow bark
Bronchitis	Garlic, hore-hound, mullein
Cancer	Pau d'arco
Colds and flu	Anise seed, echinacea, elderberry, ginger, goldenseal
Constipation	Fenugreek, flaxseed, 10 glasses of water a day, psyllium seeds, powders, and/or husks
Convulsions, possible cause of	Sage
Coughing	Coltsfoot, slip-pery elm
Cuts in the skin	Tea tree oil
Diarrhea	Fenugreek
Fever	Ginger, red pepper, cinnamon, cranberry
Fungus infections	Chamomile, garlic, cloves
Garlic odor	Parsley
Gum infections	Chamomile
Hay fever/allergy reaction Headaches, migraines	Chamomile Feverfew

Headaches, possible cause of	Licorice
High blood pressure, cause of	Ephedra, ginkgo,ginseng, guarana
High blood pressure, possible cause of	Licorice, ephedra, ginseng, ginkgo, guarana
Immune system weakness	Echinacea
Indigestion	Caraway, cloves,dill, fennel, ginger, horse balm, peppermint
Insect stings	Garlic
Laxative effect, cause of	Cascara sagrada
Liver clogging	Milk thistle
Liver damage, possible cause of	Plant oils, tea tree oil, pennyroyal, comfrey
Menstrual problems	Bilberry
Metabolism	Red pepper (cayenne)
Minor skin burns	Medicinal aloe vera pulp, Calendula (external only)
Mood enhancer, temporary remedies	Clove, licorice, passion flower, rosemary, tilleul, various flower
Poison oak, sumac, ivy	Medicinal aloe vera plant pulp
Potassium loss, possible cause of	Licorice
Prostate problems	Saw palmetto
Sinus infection	Echinacea, goldenseal, garlic
Sore throat elm	Eucalyptus, licorice, slippery
Sodium retention, possible cause of	Licorice
Soporific (sleep inducing) state, cause of	Chamomile, valerian
Sunburn	Medicinal aloe, black tea in wet tea bags, calendula ointment
Water retention, possible cause of	Licorice

| Wrinkles in the skin | Grapes (mashed) horse chest-nut, witch hazel, cocoa butter, all externally applied |

Chapter 7: Healing Supplements for Women

Alzheimer's disease	Hormone-replacement therapy, fluoride
Breast health	Selenium; vitamin E; beta-carotene (from food); (one cup of broccoli, five carrots, and three mugs of green tea daily); milk thistle; omega-3 fats; gamma linoleic acid (GLA); pure ginseng
Colon cancer	Calcium, magnesium
Fatigue	Homeopathic remedies
Insomnia	Hyland's Nerve Tonic
Memory problems	Vitamin B-complex, cobalamin, folacin, thiamin
Menopause	Hormone–replacement therapy, estral, estradiol, Estrace, Prometrium, pyridoxine, black cohosh, chaste-berry, mela tonin, homeopathic remedies
Nervous	Hyland's Nerve exhaustion Tonic
Nervousness	Homeopathic remedies
Osteoporosis	Calcium
Premenstrual symptoms	Pyridoxine, vitamin E; Dong Quai; gamma linoleic acid (GLA) in evening primrose oil and black currant oil capsules; vitus agnus castus (chaste tree, also called chasteberry); black haw bark; raspberry leaf tea; pycnogenols; kavakava root; homeopathic remedies
Tension	Homeopathic remedies

Index

Bronchitis, garlic for, 129
Burns, Dr. Eugene, 25, 29, 38-39, 43, 44
B vitamins, 26-32
  biotin (B$_7$), 27, 28, 30-31
  cobalamin (B$_{12}$), 27, 28, 29, 30, 107, 145
  folic acid (B$_9$), 28, 31-32, 107, 145
  food sources, 28, 29
  healing benefits, 27-28
  and homocysteine, 107
  how to take, 28, 30
  and memory loss, 145-46
  niacin (B$_3$), 14, 28, 29
  numbering of, 30
  pantothenic acid (B$_5$), 28, 31
  pyridoxine (B$_6$), 28, 30, 107, 111, 139
  RDAs, 23
  riboflavin (B$_2$), 28, 29
  thiamin (B$_1$), 28-29, 145

Cadmium, 54
Cadoux, Dr. Alex, 115, 140
Caffeine, 21
Calcium
  daily value amount, 55
  function and sources, 50-51, 56
  healing benefits, 56
  maximum dosage, 57, 148-49
  needs after age fifty, 57
  to prevent colon cancer, 56, 149
  to prevent osteoporosis, 148-49
Calcium lactate, 56
Calendula ointment, 124
Cancer
  of the breast, 146
  chemotherapy and radiation aids, 80
  of the colon, 149
  of the esophagus, 62
  fighting with antioxidants, 79-81
  of the mouth, 37
  and selenium levels, 65
Capsules, 15
Caraway, 125
Carminatives, 122
Carnitine
  combined with glutamine, for alcoholism, 106

combined with stimulants, dangers of, 105
  function, 97
  healing benefits, 104-5
  safe dosage, 104-5
Carotenoids, 71, 83. *See also* Beta-carotene
Cascara sagrada, 127
Cataracts, 29
Cervical dysplasia, 27-28, 30, 34
Chaitow, Dr. Leon, 106, 109
Chamomile, 125, 127
Chasteberry (vitus agnus-castus), 139-40, 143
Chelated minerals, 50
Chemotherapy, 34, 80
Children, vitamins for, 22
China, 90
Chinese medicine, 8
Chiropractors, 5-6
Chlorella, 122-23
Chloride
  food sources, 51
  function, 51
  healing benefits, 56-57
  RDA, 55
Cholesterol, lowering, 31, 82
Choline, 97
Chopra, Deepak, 116
Chromium
  function, 51
  healing benefits, 58
  maximum dose, 58
  RDA, 55
Chromium picolinate, 58
Cigarette smoking
  cadmium danger, 54
  reducing effects of, 29, 34, 41, 78
Cinnamon, 125, 127
Citronella, 125
Cloves, 120, 125
Cobalamin. *See* Vitamin B$_{12}$
Cocaine, 110
Cocoa butter, 125
Coenzyme Q10
  as an antioxidant, 25, 76, 88
  dosages, 77, 82, 88
  healing benefits, 76, 77, 82, 88
Coffee, 110
Colds, zinc for, 67-68
Colloidal minerals, 50
Colon cancer, calcium for, 149
Coltsfoot, 125
Comfrey, 126

Complete Herb Book, The (Stuckey), 129
Complete Spice Book, The (Stuckey), 129
Constipation
  algae for, 122-23
  B vitamins for, 28
Copper
  as an antioxidant, 58, 85-86
  dosage, 55, 86
  function and sources, 51
  healing benefits, 58-59
Council on Chiropractic Education, 6
Council on Homeopathic Certification, 7-8
Cranberry, 130
Creatine phosphate (CP), 108
Cysteine, 97, 105

Daily Reference Value (DRV), 15
Daily Value (DV), 15, 22
  for minerals, 55
Dartmouth College, 58
Davis, Adelle, 59
Decoctions, herbal, 119
Depression
  folic acid for, 31
  glutamine and tyrosine for, 106
  herbal mood enhancers, 119
  phenylalanine for, 109
Deva Flower Remedies, 124
Dill, 126
Diuretics, 21
Dong quai, 139
Drowsiness, biotin for, 27
Dyes, in supplement tablets, 24

Echinacea, 4, 71, 116, 126-27
Edell, Dean, 58
Elderberry, 127-28
Electrolytes, 54
Endorphines, 109
England, 7
Ephedra (ma huang), 127
Eskimo diet, 99
Essential amino acids, 96
Estrace, 142
Estradiol, 142
Estrogen, 137, 141-42
  and osteoporosis, 148
  replacement therapy, 141-42, 146, 148

Books in the
Smart Guide™ Series

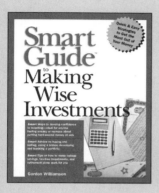

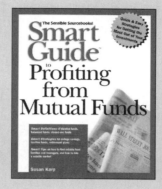

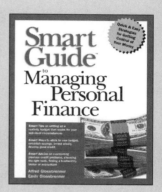

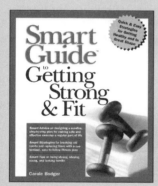

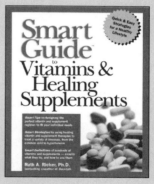

Smart Guide™ to
Getting Strong and Fit

Smart Guide™ to
Getting Thin and
Healthy

Smart Guide™ to
Making Wise
Investments

Smart Guide™ to
Managing Personal
Finance

Smart Guide™ to
Profiting from Mutual
Funds

Smart Guide™ to
Vitamins and Healing
Supplements

Available soon:

Smart Guide™ to
Boosting Your Energy

Smart Guide™ to
Healing Foods

Smart Guide™ to
Home Buying

Smart Guide™ to
Relieving Stress

Smart Guide™ to
Starting and Operating
a Small Business

Smart Guide™ to
Time Management